Pesticides

A Toxic Time Bomb in Our Midst

Marvin J. Levine

Westport, Connecticut
London

Library of Congress Cataloging-in-Publication Data

Levine, Marvin J., 1930–
     Pesticides : a toxic time bomb in our midst / Marvin J. Levine.
          p.   cm.
     Includes bibliographical references and index.
     ISBN 978–0–275–99127–2 (alk. paper)
     1. Pesticides—Health aspects.   2. Pesticides—Environmental aspects.
     3. Pesticides—Toxicology.   I. Title.
     RA1270.P4L48   2007
     363.738'4—dc22   2007000057

British Library Cataloguing in Publication Data is available.

Library of Congress Catalog Card Number: 2007000057
ISBN-13: 978-0-275-99127-2
ISBN-10: 0-275-99127-X

First published in 2007

Praeger Publishers, 88 Post Road West, Westport, CT 06881
An imprint of Greenwood Publishing Group, Inc.
www.praeger.com

Printed in the United States of America

The paper used in this book complies with the
Permanent Paper Standard issued by the National
Information Standards Organization (Z39.48–1984).

10 9 8 7 6 5 4 3 2 1

Contents

To Dina

The writing of this book came about through a serendipitous circumstance. I wrote a book dealing with child labor in the United States, published three years earlier, containing a chapter that included a description of the hazards migrant farmworkers and their children face, not the least of which was exposure to pesticides. While mentioning the dangers pesticides posed, I thought that a book on pesticides could be a future undertaking. However, I put it on the back burner and turned my full attention to the project at hand.

Approximately one year ago, I was considering other potential book topics, when my wife, Dina, suggested a book on pesticides. I recalled that pesticides had been mentioned in the child-labor book and agreed that it could be an interesting and manageable proposition. Soon, I began research on that topic.

Agriculture, covered in the second chapter, was singled out because a substantial majority, 70 percent or more, of pesticides are applied in U.S. farming operations on an annual basis. Billions of dollars are spent in the sale and use of these hazardous chemicals. Their deleterious impact upon the health of farmers, farmworkers, and their children will be examined, with special emphasis on threats to the well-being of several million migrant farmworkers' families.

Another important topic deals with the health problems attributed to pesticide residues in food, most of which is grown using a variety of pesticides. Acute and chronic effects on children's health will be investigated.

A largely overlooked area also merits consideration. Nearly 90 percent of all U.S. households use pesticides, primarily for insect control. The number and concentration of pesticides detected in the indoor air of homes is typically greater than those discovered in the air outdoors. People spend the majority of their time indoors, more than 90 percent of each day. Millions of pounds of these toxic chemicals are also applied on American lawns and gardens, when safer alternatives are available.

Furthermore, in too many of the 110,000 school districts across the nation, untrained personnel are making critical decisions day in and day out about the use of pesticides in school buildings and on school grounds. Children attend at least 180 days of school each year. An increased incidence of learning disorders has been linked to this pesticide use. Federal law permits protections for farmers from re-entering fields too soon after pesticide applications, but no such measures are available in the case of many schools.

A serious health problem is also posed by some level of pesticide contamination of drinking water in every state nationwide, in both agricultural and urban regions. Continuous monitoring will be essential to alleviate this health peril to our population.

In addition, other topics scrutinized here include testing, data collection, legislation, regulation, and political influence exerted by pesticide manufacturers.

I hope this book will heighten public awareness of the dangers pesticides pose for humans, wildlife, and the environment.

The following persons and organizations deserve thanks for their assistance: Linda Greer of the Natural Resources Defense Council in Washington, D.C.; Carol Raffensperger and Ted Schettler of the Science and Environmental Health Network in Ames, Iowa; The Center for Health, Environment & Justice in Falls Church, Virginia; Aviva Glazer of the School Pesticide Monitor in Washington, D.C.; Beyond Pesticides in Washington, D.C.; and Suzanne and Ralph Tarica.

A special debt is owed to my editors at Praeger Publishers, Hilary Claggett, and James R. Dunton, for their timely assistance in the preparation of the manuscript.

Last, but not least, I take full responsibility for any errors of omission or commission.

One

The Pesticide Problem

If we don't change direction soon, we'll end up where we're going.
                                        —Professor Irwin Corey

Professor Irwin Corey was considered a guru of comedy by those who remember him when he was a regular on the Steve Allen television shows of the 1950s and 1960s. However, there is nothing humorous about the subject of this book—pesticides—and his message has become increasingly relevant.

There is growing public concern regarding pesticide exposure, and for good reason. Studies have shown that all people, especially children, pregnant women, farmers, farmworkers, and the elderly, may experience negative health effects from exposure to pesticides. Pesticide exposure can cause acute poisoning, cancer, neurological damage, birth defects, and reproductive and developmental harm.[1] Much evidence has revealed that many commonly used pesticides can suppress the normal response of the human immune system, making the body more vulnerable to invading viruses, bacteria, parasites, and tumors, increasing the incidence of disease and some cancers.[2] Some evidence indicates that pesticides may reduce male sperm counts.[3] Unfortunately, pesticides are widely used in our environment to control pests, but we the people rarely hear about it. Fortunately, there are ways to reduce pesticide use and exposure.

Pesticides, by design, are toxic to certain life forms. Currently in the United States there are more than 17,000 registered pesticide products and more than 800 active ingredients. Acute pesticide-related illness and injury continues to be a problem. According to poison control center data, there are approximately 18,000 unintentional pesticide exposures each year. Approximately 1,400 of these are occupational.[4] According to Bureau of Labor Statistics data, annually there are 500 to 900 lost work-time illnesses caused by pesticide exposure. Finally, there are approximately fifteen to twenty death certificates per year that contain codes for unintentional pesticide poisoning. All of these estimates are thought to be underestimates of the true incidence of unintentional acute pesticide-related illness and injury.[5]

Historical Patterns of Pesticide Use

The era of pesticides began in the nineteenth century when sulfur compounds were developed as fungicides. In the late nineteenth century, arsenic compounds were introduced to control insects that attack fruit and vegetable crops; for example, lead arsenate was used widely on apples and grapes. These substances were acutely toxic. In the 1940s the chlorinated hydrocarbon pesticides, most notably DDT (dichlorodi-phenyltrichloroethane), were introduced. DDT and similar chemicals were used extensively in agriculture and in the control of malaria and other insect-borne diseases. Because they had little or no immediate toxicity, they were widely hailed and initially believed to be safe.[6]

Widespread use of synthetic pesticides in the United States began after World War II. The ingredients for many of today's pesticides were, in fact, created as weapons of war.[7] Before the development of synthetic pesticides, farmers used naturally occurring substances such as arsenic and pyrethrum.[8] Pesticide use was credited with increasing crop yields by reducing natural threats and became an integral part of agricultural practices by the mid-1950s. Over the past five decades, American agriculture has dispersed thirty billion pounds of pesticides into the environment.[9] Also, beginning in the late 1940s, federal and local governments sponsored the widespread spraying of DDT and other chemicals in urban communities in an effort to eradicate mosquitoes, fire ants, gypsy moths, the Japanese beetle, and other insects judged to be harmful.

Every year in the United States, 1.1 billion pounds of active pesticide ingredients are released into the environment; 834 million pounds (77 percent) are used in agriculture, the remainder for non-agricultural purposes. If the use of wood preservatives, disinfectants, and sulfur is included, the yearly amount of pesticide usage increases to 2.2 billion pounds of active ingredients.[10] Altogether, U.S. pesticide usage equals more than four pounds per person annually.[11]

Insects, however, quickly develop resistance to pesticides. In addition, broad-spectrum pesticides kill natural predators that keep pests in check. Use of synthetic pesticides—including insecticides, rodenticides, fungicides, herbicides, and others—has increased more than thirty-three-fold in the last half century. Ironically, it is estimated that more of the U.S. food supply is lost to pests today (37 percent) than in the 1940s (31 percent). Total crop losses from insect damage alone have nearly doubled from 7 percent to 13 percent during that period. Cultivation of four crops—soybeans, wheat, cotton, and corn—consumes around 75 percent of the pesticides used in the United States.[12]

Following World War II, pesticides were a component of what was predicted to be a "green revolution" of abundant food for the world. Over the past fifty years, agricultural production in many areas of the world has increased dramatically, partly because of the use of herbicides and insecticides. Health benefits, such as those related to the eradication of malaria-carrying mosquitoes, were also foreseen and, in many cases, attained.

In May 1962, biologist Rachel Carson alerted the public to the side effects of pesticides in her book, *Silent Spring*. Questions were raised about the actual (rather than the perceived) benefits of pesticides, along with questions about environmental and public health risks.[13]

The pathways of human exposure to pesticides are numerous. Pesticide residues are found virtually everywhere: in the office and home, on food, in drinking water, and in the air.[14] Throughout more than a half century of pesticide use, most pesticides have never been systematically reviewed to evaluate their full range of long-term health effects on humans, such as potential damage to the nervous, endocrine, or immune systems. The Environmental Protection Agency (EPA) considers only cancers in determining the potential threat of pesticides to human health. Until recently, cancer has been considered the most sensitive end point—if you could prevent cancer, you could prevent other chronic diseases. Furthermore, scientists have been able to develop the model by which they can extrapolate cancer data from animal studies. The concept that cancer is the most sensitive end point is now being seriously questioned. The effects of pesticides on wildlife are also not well documented. It wasn't until 1985 that the EPA reviewed an insecticide solely on the basis of its effects on wildlife.[15] Since then, the EPA has banned some pesticides based partially on their effects on the environment and wildlife. Discoveries of pesticide residues have also resulted in fishing bans in bays, lakes, and rivers.[16]

Agricultural pesticides have prevented pest damage of between 5 percent and 30 percent of potential production in many crops.[17] Pesticides, however, have posed a number of problems for agriculture, including the killing of beneficial insects, secondary pest outbreaks, and the development of pesticide-resistant pests.[18] Several studies have shown a decrease in the effectiveness of pesticides. According to one study, 7 percent of U.S. agricultural production was lost to pests in the 1950s; in 1993, 13 percent of all production was lost to pests.[19] A different study concluded that crop losses from pests increased from 30 percent in 1945 to 37 percent in 1990. During that same period farmers used thirty-three times more pesticides.[20]

Today, 440 species of insects and mites and more than seventy fungi are now resistant to some pesticides.[21] Consequently, it has become necessary to use larger doses and more frequent applications of pesticides. Combining pesticides, or substituting more expensive, toxic, or ecologically hazardous pesticides, occurs more frequently. In addition to the problem of pesticide resistance, millions of dollars worth of crops have been lost as a result of improper pesticide application.[22]

Health Effects on Children

Pesticides have been associated with the development of certain cancers in children, including leukemia, sarcomas, and brain tumors. Many classes of pesticides have been shown to adversely affect the developing nervous system of animals used in experiments. Parental exposure to pesticides has been linked with birth defects in children. New studies suggest that pesticides may compromise the immune systems of infants and children. Children are exposed to pesticides at home, at school, in playgrounds and parks, in food, and in water. Nationwide, 85 percent of households had at least one pesticide, and 47 percent of households with children under the age of five were found to store at least one pesticide within the reach of children. Parents can eliminate the use of pesticides in and around their homes and workplaces and pressure school boards to reduce pesticide use in schools. If possible, parents can buy organically grown and in-season foods. Congress passed legislation in 1996 designed to improve regulation of pesticides, particularly in food, so that children are adequately protected. The implementation of this law will be a critical test of the EPA's intention to safeguard the next generation. Additional reforms needed include reducing the use of pesticides, better testing of pesticides' ability to affect infants and young children, and more data on children's exposure to pesticides.[23]

Controversy has arisen regarding the apparent increase in the incidence of childhood cancer in the United States. Some investigators, particularly at the EPA, have raised concerns that this increase may reflect new or increasing environmental exposures. The alternative view is that there has been little overt change in incidence, and that apparent increases in, for example, brain tumors, reflect changes in medical practice and diagnostic methods rather than a true increase in occurrence. Part of the difficulty in understanding childhood cancer trends lies in the relative rarity of most cancer types and the lack of a national system of cancer registration that would enable researchers to track incidence on a nationwide scale.[24]

Children may be more susceptible than adults to environmental health risks because of their physiology and behavior. They eat and drink more and breathe more air in proportion to their body weight than adults. They also play close to the ground and put objects in their mouths. Their bodies are still developing, and they may be less able than adults to metabolize and excrete pollutants.

In 1996, poison control centers nationwide were notified about approximately 80,000 children (aged from birth through nineteen) who were exposed to common household pesticides, an estimated one-quarter of whom developed symptoms of pesticide poisoning. From 1992 to 1998, an estimated 24,000 emergency room visits resulted annually from pesticide exposure; 61 percent of the cases involved children younger than age five.[25]

New Discoveries About Pesticides

Although pesticides do offer certain benefits for farmers and others, new scientific research is revealing some important health-related issues associated with their usage. Recently, for example, some scientists have become convinced that there is a relationship between pesticides that mimic the estrogen hormone and the disruption of the endocrine systems in humans and wildlife. This potentially could contribute to serious health problems, including breast and other types of cancer in humans, and reproductive disorders.[26] Currently, in registering pesticides, the EPA does not require tests for estrogen involvement; if a pesticide is found to be estrogenic, the EPA has no method of removing it from the market.[27]

Though there is no conclusive evidence to date, several studies have indicated that chemicals that imitate estrogen might cause reproductive problems in animals. For example, one study found that male alligators exposed to pesticides in Florida are having difficulty reproducing, partly because their penises are not developing to normal size. This reproductive interference could be related to exposure to estrogenic pesticides. It also has been reported that some birds, fish, amphibians, and mammals are being "feminized" by exposure to low levels of pesticides and other industrial chemicals.[28]

Pesticide Usage

Pesticides of various types are used in most sectors of the U.S. economy. In general terms, a pesticide is any agent used to kill or control undesired insects, weeds, rodents, fungi, bacteria, or other organisms. Thus, the term "pesticides" includes insecticides, herbicides, rodenticides, fungicides, nematicides, and acaricides as well as disinfectants, fumigants, wood preservatives, and plant growth regulators. Pesticides play a vital role in controlling agricultural, industrial, home/garden, and public health pests. Many crops, commodities, and services in the United States could not be supplied in an economic fashion without controlling pests using chemicals or other means. As a result, goods and services can be supplied at lower costs and/or with better quality. As has been pointed out, these economic benefits from pesticide use are not achieved without potential risks to human health and the environment due to the toxicity of pesticide chemicals. For this reason, these chemicals are regulated under federal or state pesticide laws to avoid unacceptable risks.

The EPA registers pesticides for use and requires manufacturers to label pesticides about when and how to use them. It is important to remember that the "cide" in pesticide means "to kill." These products can be dangerous if not used properly.

Annual pesticide use in the United States equals about 8.8 pounds per capita, relatively stable at roughly 2.2 billion pounds of active ingredients, according to an EPA pesticide industry sales and usage report. According to the report, use of what are considered "conventional pesticides" remains at about 1.1 billion pounds of active ingredients, but the addition of wood preservatives and disinfectants pushes total pesticide use to about 2.2 billion pounds of active ingredients. Pesticides are used on more than 900,000 U.S. farms and in 69 million households, the report indicated, while the herbicides atrazine and metolachlor are the two most widely used pesticides in the country, at 70 million to 75 million pounds and 60 million to 65 million pounds, respectively.[29]

Three Major Groups of Conventional Pesticides

The first group consists of chlorinated hydrocarbons, also known as organochlorines. These pesticides generally break down very slowly and can remain in the environment for long periods of time. Dieldrin, chlordane, aldrin, DDT, and heptachlor are pesticides of this type. The second group is known as organic phosphates or organophosphates. These pesticides are often highly toxic to humans, but generally do not remain in the environment for long. Diazinon, malathion, dimethoate, and chlorpyrifos are pesticides of this classification. The last group is the carbamates. They are generally less toxic to humans, but concerns persist about the potential effects of some carbamates on immune and central nervous systems. Carbaryl, carbofuran, and methomyl are examples of carbamates.[30]

Pesticide Safety Myths

There is no such thing as a "safe" pesticide. In fact, pesticide labels describe their products as possessing varying degrees of toxicity. For that matter, it is illegal for pesticide manufacturers to allege safety as a pesticide characteristic in their promotional efforts. Different pesticides affect people in different ways. Some cause cancer and are listed as "known" or "possible" carcinogens as identified by the EPA or state environmental agencies. Some are nerve toxins, which affect the enzyme responsible for the basic operation of the brain and nervous system. Many originate from World War II research on chemical weapons. These include organophosphate and carbamate insecticides such as chlorpyrifos and diazinon. Acute (immediate) poisoning symptoms are flu-like, featuring nausea, vomiting, diarrhea, or dizziness. These pesticides may also impair memory, learning ability, ability to focus, and even normal behavior. Reproductive and developmental toxins are those that impact the development of children.[31] Exposure to these chemicals may jeopardize a child's mental or physical development. Pregnant women exposed to these chemicals may face increased risk of birth defects in their unborn children. Hormone-mimicking toxins also known as endocrine disruptors can disrupt delicate hormonal processes in wildlife and humans. Hormones act as chemicals in the human body, triggering a wide array of biological processes. They can impact height and weight, gender differentiation, the development of reproductive organs, and energy levels. Because hormones function at very low levels, these pesticides can have dramatic effects even at modest levels of exposure.[32]

Pesticide Resistance

In addition to directly poisoning our environment and our food, pesticides pose a serious threat to our food production system itself. From one viewpoint, pesticides are wonder chemicals that have increased food production by 20 percent since 1940 by reducing pest damage. Yet over the same period, they have also created at least 261 strains of insect species, sixty-seven strains of plant pathogens, two strains of nematodes (parasitic worms), and four (or by some counts, nineteen) strains of weeds that they cannot kill. While insecticide use has increased tenfold since the 1940s, crop losses to insects doubled.[33]

The key to this paradox is the selection for resistance that pesticides exert on their target pests. Pesticides never kill 100 percent of a pest population, and the survivors tend to have a lower susceptibility to that particular chemical. With every repeated application of the same pesticide, these naturally resistant individuals make up a higher percentage of the population, until a highly resistant strain of pest evolves. When the conditions are right, the pesticide kills a large percentage of the pest population, the pest completes several life cycles per year, and little movement from untreated populations occurs. Then resistance can develop very rapidly.

Resistance to one pesticide often confers resistance, or faster development of resistance, to a whole family of related pesticides. Alternating different pesticides or applying a mixture of chemicals can sometimes delay the development of resistance, but it can also promote the development of super-resistant pests, called superpests, which are resistant to multiple pesticides. Superpests have already developed and threaten a number of crops throughout the world.[34]

Fate of Pesticides in the Environment

Ideally, a pesticide stays in the treated area long enough to produce the desired effect and then breaks down into harmless materials. Three primary modes of degradation occur in soils:

- biological—breakdown by micro-organisms

- chemical—breakdown by chemical reactions, such as hydrolysis (soluble decomposition) and oxidation

- photochemical—breakdown by ultraviolet or visible light

The rate at which a chemical degrades is expressed as the half-life, which is the amount of time it takes for half of the pesticide to be converted into something else, or until its concentration is half of its initial level. The half-life of a pesticide depends on the soil type, its formulation, and environmental conditions such as temperature and moisture levels. Other processes that influence the fate of the chemical include plant absorption, soil adhesion, leaching, and vaporization. If pesticides migrate from their targets due to wind drift, runoff, or leaching, they are considered to be pollutants. The potential for pesticides to move depends on the chemical properties and formulation of the pesticide, soil properties, the rate and method of application, pesticide persistence, frequency and timing of rainfall, irrigation, and depth to ground water.[35]

Pesticide Toxicity

Toxicity is the inherent ability of a pesticide to cause injury or death, indicating how poisonous the chemical is. Acute toxicity is the ability of a substance to cause harm as the result of a single dose or exposure to a chemical. Chronic toxicity is the ability of a substance to cause injury as the result of repeated doses or exposures over time. Any chemical substance is toxic if it is ingested or absorbed in excessive amounts. Table salt, for example, if consumed in excess, can be toxic. The degree of danger or hazard when using a pesticide is determined by multiplying toxicity times exposure.

The designation given to a pesticide indicating its relative level of toxicity is called the lethal dose, or LD_{50} value. This value identifies the dosage necessary to kill 50 percent of a test population. The lethal dose is expressed in milligrams of chemical per kilogram of body weight of the test population. The lower the LD number, the more toxic the material. The toxicity rating is important as an indicator, but the length of exposure, type of exposure, and other factors also impact the relative hazard of any pesticide. The toxicity of pesticides is often measured using an LD_{50} (lethal dose) or an LC_{50} (lethal concentration). Both the LD_{50} and LC_{50} measure only acute effects and therefore provide no information about a chemical's connection to long-term health issues.[36]

The tests for acute and chronic toxicity are the only science-based methods currently used to predict risks to users and consumers. But they have limitations. These tests are usually done on rodents, which may not always accurately predict effects on humans. Plus, they do not take into consideration possible interactions and consequences of several compounds acting together.

All labels include the warning, "Keep out of reach of children." In addition, most labels include "signal words" which give an indication of the pesticide's toxicity or corrosiveness. These signal words are relative terms. They indicate how pesticides compare to one another. Even if a pesticide is considered to be relatively low in toxicity, it can be a deadly poison at a fairly low dose.[37]

Inert Ingredients

Pesticide products contain both "active" and "inert" ingredients. These terms have been defined by a federal law, the Federal Insecticide, Fungicide, and Rodenticide Act (FIFRA) of 1947. An active ingredient is one that prevents, destroys, repels, or mitigates a pest, or is a plant regulator, defoliant, desiccant, or nitrogen stabilizer. By law, the active ingredient must be identified by name on the label together with its percentage by weight. An inert ingredient is simply an ingredient in the product that is not intended to affect a target pest. For example, isopropyl alcohol may be an active ingredient and antimicrobial pesticide in some products; however, in other products, it functions as a solvent and may be considered inert. The law does not require inert ingredients to be identified by name and percentage on labels, but the total percentage of such ingredients must be declared.[38]

Inert Name Change

In September 1997, the EPA issued a regulation notice to encourage manufacturers, formulators, producers, and registrants of pesticide products to voluntarily substitute the term "other ingredients" as a heading for the "inert" ingredients in the ingredient statement on pesticide labels. The EPA made this change after learning the results of a consumer survey on the use of household pesticides. Many comments from the public and the consumer interviews prompted the EPA to discontinue the use of the term "inert." Many consumers are misled by the term "inert ingredient," believing it to mean "harmless." Since neither federal law nor the regulations define the term "inert" on the basis of toxicity, hazard, or risk to humans, non-target species, or the environment, it should not be assumed that all inert ingredients are non-toxic.[39]

Status of Inert Ingredients

Inert ingredients have definitely not been given a clean bill of health. For example, it is not clear which components of weed killers are carcinogens. The question revolves around whether it is the active ingredients, the dioxins, that contaminate the active ingredients during manufacture, or the inert ingredients, which frequently constitute 90 to 99 percent of pesticides. Inert ingredients are added as fillers or to give the pesticide a desirable quality. The EPA lists 2,000 chemicals that have been approved for use as inert ingredients. These include urea formaldehyde, carbon tetrachloride (known to cause cancer), chloroform (also a known carcinogen), toluene, xylene, cadmium, and lead compounds. Pesticide manufacturers have successfully claimed that the components of inert ingredients are trade secrets not required to be disclosed to potential competitors. Furthermore, federal law imposes a $10,000 penalty on any employee who reveals the contents of inert ingredients in pesticides.[40]

Agricultural Pesticides

Much of modern farming relies on pesticides to produce food of a high quality and ensure consistent supplies. In some cases pesticides can make the difference between success and failure of a crop. Pesticides are a vital part of modern agriculture, protecting food and fiber from damage by insects, weeds, diseases, and rodents. U.S. agriculture companies spend about eight billion dollars annually on pesticides, which accounts for more than 70 percent of all pesticides sold in the country.[41] It is estimated that each dollar invested in pesticide control returns approximately four dollars in crops saved from pests. Farmers' expenditures on pesticides are about 4 to 5 percent of total farm production costs.[42]

The dependence of agriculture on chemical pesticides developed over the last sixty years as the agricultural sector shifted from labor-intensive production methods to more capital- and chemical-intensive production methods. Sixty years ago, most crops were produced largely without the use of chemicals. Insects and weeds were controlled by crop rotations, destruction of crop refuse, timing of planting dates to avoid high pest population periods, mechanical weed control, and other farming practices. While these practices are still in use, changes in technology, costs, and government policies have led to the development of today's chemically intensive farming methods.

Usage of conventional pesticides on farms in the United States increased from about 400 million pounds (of active ingredients) in the 1960s to more than 800 million pounds in the late 1970s and early 1980s, primarily due to the widespread adoption of herbicides in corn production. Since that time, usage has been somewhat lower, ranging from about 700 to 800 million pounds annually.[43] Pesticide usage in agriculture can vary considerably from year to year, depending on weather, pest outbreaks, crop acreage, and economic factors such as pesticide costs and crop prices. Whereas the quantity of pesticides used by farmers has fallen off slightly in recent years, total expenditures on pesticides are still increasing.

During the 1960s, agricultural pesticide use was dominated by insecticides, accounting for about half of all pesticides used. The quantity of insecticides applied fell as the organochlorines (DDT, aldrin, and toxaphene) were replaced by pyrethroids and other chemicals that require lower application rates. Today, 70 percent of pesticides used are herbicides, with corn leading all other crops by a substantial margin in total pesticide use. Rice, potatoes, vegetables, and fruits, however, actually use pesticides more intensively than corn and other crops. Minimum tillage practices are being adopted by many farmers, reducing the need for machinery, labor, and energy inputs, but increasing farming's dependency on pesticides even more. Pesticide use trends can vary markedly from one part of the country to another as farmers respond to local pest problems and as crop production patterns vary.

Concerns about potential risks to health and the environment resulted in amendments to FIFRA in 1972, increasing the stringency of health and safety data required to support a pesticide registration. The EPA first banned the usage of some organochlorine pesticides for agricultural purposes in the 1970s, and has since imposed use limitations on many other pesticides. The amendments also required that all existing pesticides be reregistered using current health and environmental standards. Chemical companies have responded to these regulatory pressures by marketing new chemicals that are thought to be less harmful to humans and the environment, or less likely to migrate from farm fields to contaminate groundwater and surface water.

Schools and Pesticides

Safeguarding children's health while at school is a priority for parents, teachers, school administrators, lawmakers, and clinicians. Yet children are continually and unknowingly exposed to toxic chemicals while in and around school buildings. Substantial scientific evidence indicates that children are at risk for diseases as a result of these exposures.

Despite the hazards to children and the environment, pesticides have become a preferred approach to controlling pest problems in many schools and school districts. Toxic chemicals are being used on school athletic fields, shrub beds, parking lots, tracks, play areas, and in cafeterias, classrooms, gymnasiums, and restrooms. Too often pesticides are applied by unlicensed personnel, or applied on a calendar basis whether pests are present or not.

In general, research demonstrates that pesticide poisoning can lead to poor performance on tests involving intellectual functioning, academic skills, abstract reasoning, flexibility of thought, and motor skills. Other areas affected include memory disturbances and inability to focus attention, reduced perceptual speed, and deficits in intelligence, reaction time, and manual dexterity. Increased anxiety and emotional problems have also been reported.

Pesticide opponents estimate there are some fifty insecticides, herbicides, and fungicides commonly used in and around schools. Some are implicated in reproductive and neurological problems, kidney and liver damage, and cancer. Additionally, the following have been reported as adverse health effects of forty-eight commonly used pesticides in schools: twenty-two are probable or possible carcinogens, twenty-six have been shown to have reproductive effects, thirty-one damage the nervous system, thirty-one injure the liver or kidneys, forty-one are sensitizers or irritants, and sixteen can cause birth defects. Because most of the symptoms of pesticide exposure, from respiratory distress to difficulty in concentration, are common in schoolchildren and may also have other causes, pesticide-related illnesses often go unrecognized and unreported.[44]

The GAO Study

In the fall of 1999, the General Accounting Office (GAO), at the request of Democratic Senator Joseph Lieberman of Connecticut, conducted a national review of the extent to which pesticides are used in and around the nation's 110,000 public schools and the magnitude of the risk of exposure to children. The report found that the data on the amount of pesticides used in the nation's public schools is neither available nor collected by the federal and most state governments. The study also revealed that the EPA is not doing enough to protect children from pesticides, and that there is limited information on how many children are exposed to pesticides in schools. The GAO cited the EPA's analysis of the American Association of Poison Control Centers' Toxic Exposure Surveillance System, documenting 2,300 school pesticide exposures from 1993 to 1996. Because most of the symptoms of pesticide exposure, from respiratory distress to difficulty in concentration, are common and may be assumed to have other causes, it is suspected that pesticide-related illnesses are much more prevalent than presently indicated.

Specifically, the GAO found that:

1. There are no comprehensive, readily available national or state-by-state data on the amount and kinds of pesticides being used in schools today.

2. Although FIFRA requires pest control companies to keep records for two years on the amount and site of pesticide applications, only one state requires them to report this information to the relevant agency.

3. There is little information available about illnesses related to pesticide exposure. The GAO documented 2,300 cases of exposure at schools from 1993 to 1996, but noted that this information is incomplete and unreliable because of the lack of record-keeping, and therefore likely underestimates how often children are exposed. In addition, of those 2,300 cases, the outcomes in 1,000 of them are not known, or more than 40 percent are incomplete. For the cases where follow-up did occur, 329 individuals were seen at health care facilities, fifteen were hospitalized, and four were treated at intensive care units.

4. Eight states collect information on the use of pesticides within their states, but only two collect information on pesticides used in schools. No state collects information on exposure patterns in schools.

5. There are no standard criteria for clearly identifying illnesses linked to pesticide exposure; misclassification of pesticide illness is common.[45]

Eliminating pesticides from the school environment is critical to lowering children's total exposure. Children spend an average of six to seven hours per day, five days per week, 180 days per year, in school. The only other place where children spend more time is in their homes. In order to protect children's health wherever they work and play, pesticide use in schools must be reduced, and families must be routinely notified whenever pesticides will be applied in schools.

As the public becomes more aware of the health and environmental risks pesticides may pose, interest in seeking the use of equally effective alternative pest control methods increases. School administrators and others who have pest control decision-making responsibilities for school buildings and grounds should become aware of the pest control options available to them. It is in everyone's best interest to reduce exposure to potentially harmful chemicals in the educational environment.

Pesticides and Water Quality

Pesticides not absorbed by plants and soils or broken down by sunlight, soil organisms, or chemical reactions may ultimately reach groundwater sources of drinking water. This depends on the nature of the soil, depth to groundwater, chemical properties of the pesticide, and the amount and timing of precipitation or irrigation in an area. Usually, the faster a pesticide moves through the ground, as with sandy soils and heavy rainfall or irrigation, the less filtration or breakdown. Heavier soils, combined with lower moisture levels and warmer temperatures, provide a greater opportunity for pesticides to break down before reaching groundwater. The amount of a pesticide detected in well samples also relates to the kind of pesticide and the amount originally applied. Contamination problems can result from using high concentrations of water-soluble pesticides for a specific crop in a vulnerable area.

Pesticides are, of course, designed to be toxic for certain insects, animals, plants, or fungi. But when used without regard to site characteristics, such as adsorption capacity of the soil ("adhesion"), solubility, climatic conditions, and irrigation patterns, a given pesticide can create greater environmental problems than the damage the target pest could cause. Once in groundwater, pesticides continue to break down, but usually much slower than in surface layers of soil. Groundwater carrying pesticides away from the original point of application can lead to contaminated well samples years later in a different location.

To avoid pesticide contamination, informed and careful pest control is necessary. Overapplication is a possible cause of pesticides in water supplies. Consequently, pesticides should not be viewed as the only answer to a pest problem; other methods may be appropriate. Integrated Pest Management (IPM) may include crop rotation, biological control, and soil analysis and conditioning.[46]

Health Effects

When pesticides are found in water supplies, they normally are not present in high enough concentrations to cause acute health effects such as chemical burns, nausea, or convulsion. Instead, they typically occur in trace levels, and the concern is primarily for their potential to cause chronic health problems. To estimate chronic toxicity, laboratory animals are exposed to lower-than-lethal concentrations for extended periods of time. Measurements are made of the incidence of cancer, birth defects, genetic mutations, or other problems such as damage to the liver or the central nervous system.

Although we may encounter many toxic substances in our daily lives, in low enough concentrations they do not impair our health. Caffeine, for example, is regularly consumed in coffee, tea, chocolate, and soft drinks. Although the amount of caffeine consumed in a normal diet does not cause illness, just fifty times this amount is sufficient to kill a human. Similarly, the oxalic acid found in rhubarb and spinach is harmless at low concentrations found in these foods, but will lead to kidney damage or death at higher doses.

Laboratory measurements of a pesticide's toxicity must be interpreted in the context of its potential hazard under actual field conditions. Pesticides by definition are toxic to at least some forms of life, but whether or not a particular pesticide in groundwater is hazardous to human health depends on its concentration, how much is absorbed from water or other sources, the duration of exposure to the chemical, and how quickly the compound is metabolized and excreted from the body. Drinking-water guidelines are aimed at keeping pesticides at levels below those that are considered to cause any health effects in humans. They are derived from laboratory data using one of two methods, depending on whether or not the compound causes cancer.[47]

Pesticide contamination of groundwater is a national issue because of the widespread use of pesticides, the expense and difficulty of cleansing groundwater, and the fact that groundwater is used for drinking water by about 50 percent of Americans. Concern about pesticides in groundwater is especially acute in rural agricultural areas, where more than 95 percent of the population relies on groundwater for their drinking water, although application rates and the variety of pesticides used may be greater in urban areas. Weed killers, bug killers, and other pesticides still contaminate thousands of water supplies nationwide. For hundreds of Midwestern communities, pesticide runoff to rivers and streams results in tap water commonly contaminated with five or more weed killers during peak runoff each spring and summer. Communities that use reservoirs are exposed to these mixtures year-round. Everyone who drinks the water is affected, including millions of babies who consume pesticides when parents feed them infant formula reconstituted with tap water. The EPA's review of the pesticide that most commonly contaminates tap water—the carcinogenic weed killer atrazine—has stalled, despite the fact that it contaminates some 1,500 water systems in twenty states, from New York to Hawaii, and has been banned in many European countries. Most efforts to reduce levels of weed killers in tap water have come literally at the end of the pipe; clean-up actions are instituted by local water suppliers and paid for by their customers.[48]

Caveats and Uncertainty

Pesticides are mostly modern chemicals. There are many hundreds of these compounds, and extensive tests and studies of their effects on humans have not been completed. That leads us to ask just how concerned we should be about their presence in our drinking water. Certainly, it would be wise to treat pesticides as potentially dangerous and, thus, to handle them with care. We can say they pose a potential danger if they are consumed in large quantities, but as any experienced scientist knows, you cannot draw factual conclusions unless scientific tests have been done. Some pesticides have had a designated Maximum Contaminant Level (MCL) in drinking water set by the EPA, but many have not. Also, the effects of combining more than one pesticide in drinking water might be different than those of each individual pesticide alone. It is another situation where we don't have sufficient scientific data to draw reliable conclusions.[49]

Federal Pesticide Regulation

In 1947, Congress took its first step to regulate pesticides with the enactment of FIFRA. This early statute was intended primarily to protect farmers and others from mislabeled, ineffective, or adulterated pesticides. That original document was only thirty-five pages long. By 1994, with billions of dollars on the line and as questions of possible adverse health effects and environmental impacts had been raised, FIFRA was expanded to more than 200 pages.[50] FIFRA initially granted jurisdiction over pesticides to the U.S. Department of Agriculture (the USDA), but in 1970, amid reports of the USDA's mismanagement and conflicts of interest, Congress shifted authority for pesticide regulation to the newly created EPA.[51] The USDA, however, continues to play a role: it is responsible for monitoring pesticide residues and it continues to promote the use of pesticides.

Key provisions of the current version of FIFRA include the following:

1. The EPA is responsible for setting most standards for pesticide use. States handle enforcement. The EPA has the authority to prohibit the use of a pesticide in the United States, to restrict uses, and to set the level of pesticide residues allowed on raw food.

2. Manufacturers of a chemical that is to be used to kill any pest must obtain a registration for the product from the EPA. If the pesticide is considered too dangerous to be used by the general population, the EPA may register the pesticide as "restricted use," which means it may be applied only by certified applicators or under the supervision of a certified applicator.

3. In order to obtain a registration, the pesticide manufacturer must provide the EPA with studies designed to ascertain the probable adverse effects of the pesticide on humans. The series of tests for a typical pesticide can cost the manufacturer millions of dollars.

4. The manufacturer must demonstrate that the pesticide "will perform its intended function without unreasonable adverse effects on public health or the environment," which is defined as "any unreasonable risk to man or the environment, taking into account the economic, social, and environmental costs and benefits of the use of any pesticide."[52]

5. Pesticides cannot be registered for use on food crops or animal feed until the EPA has determined residue tolerance levels—maximum allowable residues of chemicals. The Food and Drug Administration's (FDA) Federal Food, Drug, and Cosmetic Act provides authority for setting and monitoring tolerance levels. The FDA monitors residues and the EPA sets tolerance levels. In requesting tolerance levels, the manufacturer must provide the EPA with health data. These are usually drawn from animal studies conducted by the manufacturer.[53]

State Pesticide Regulations

State laws typically supplement or duplicate federal laws; however, in some instances, state laws are stricter than federal laws. Consequently, compliance with state laws often assures compliance with federal laws. Both federal and state laws provide for criminal prosecution and can impose penalties such as fines or imprisonment. In addition, common-law actions such as lawsuits also influence pesticide use. Common-law actions are for civil wrongs. Such actions are initiated by those who have suffered injury, or whose property has been damaged as the result of the acts or omissions of the pesticide user.

Under FIFRA and other federal regulations governing pesticide use, state agencies are authorized to 1) implement enforcement of federal regulations, and 2) assume responsibility for training and monitoring pesticide applicators. Additionally, state agencies enforce state laws regulating the sale and distribution of pesticides. Most state registration laws are limited to the collection of a fee to allow the sale of a pesticide product in the state, assuming the manufacturer has obtained an approval label under federal registration standards. In some states, pesticide laws may exceed the minimum standards prescribed under federal law, and additional review of pesticide products is required before use in the state is approved. In the case of pesticide applicator certification, some states implement minimum standards that are required by the EPA, while other states implement standards of certification that exceed federal standards.

In California, a unique system of pesticide laws regulates the use of state-designated, restricted-use pesticides. In this system, site-specific permits must be obtained to apply restricted pesticides and recommendations for such treatments can be made only by licensed pest control advisors.

In summary, the implementation and enforcement of pesticide regulations may differ from one state to another despite the fact that all states basically enforce their interpretation of the federal regulations established under FIFRA.[54]

Information Needs for Pesticide Registration

As we have seen, pesticide registration does not guarantee safety. Nor can anyone give that assurance. All pesticides are associated with some risk of harm to human health or the environment. Scientists and regulators know too little about pesticides and people's exposure to pesticides to offer assurances about pesticide safety. The EPA is mandated by federal law to evaluate the benefits of using a pesticide versus the risks it might pose to public health and the environment. To evaluate the risks and benefits of pesticide use, the EPA requires all pesticide manufacturers to conduct extensive scientific testing prior to product registration for sale and use in the United States. The manufacturers of all pesticides must compile and document information related to chemistry, toxicology, food residues, application rates, environmental impact assessment, and human safety. Normally, it takes five to ten years and upwards of $100 million to bring a new active ingredient (pesticide) to the point of approval for use by the public—a significant investment. But such scientific evaluation and regulatory scrutiny are essential to provide today's consumers with the benefits of high-quality food.

Registration is not a consumer product safety program. When the EPA registers a pesticide, it determines, among other things, how the pesticide must be used to minimize any risks, and this information must be printed on the label. Registration is a balancing act between a pesticide's benefits and its accompanying risks. Many pesticides used today were registered with the EPA before pesticide testing requirements were strengthened by Congress in 1978. As a result, many pesticides have not been subjected to the full range of tests currently required for new products. The EPA is now reviewing these products, and requiring additional testing, in a reregistration process that will not be completed for years to come. In the meantime, products registered under the earlier, less-stringent guidelines remain on the market and in use.

In its labyrinthine complexity, the pesticide registration process approximates an amusement park hall of mirrors, full of twists and turns, in which a potential registrant often retraces its steps and seems to be always going in circles. The EPA often calls for redundant health and environmental effects testing, studies to corroborate earlier studies, and environmental effects testing on specific ecosystems. The regulatory definition of what constitutes an adequate test is sufficiently ill defined to allow the EPA to challenge findings of studies even after the agency has approved the test protocol and the laboratory where the research is being done.[55]

Inadequate Legal Enforcement

The misuse of pesticides is responsible for dozens of deaths and hundreds of poisonings nationwide every year. Each year, poison control centers across the country receive thousands of calls involving both agricultural and nonagricultural pesticide exposure by children under six years of age. Pesticides also contribute to long-term health problems in the people and communities who are exposed. Although laws have been passed to address pesticide misuse, those laws are often ignored or underenforced.

Today, individuals or businesses who violate pesticide laws frequently escape with tiny fines, or without paying any penalty at all. Unless wrongdoers receive meaningful fines, they have little incentive to comply with the law, and implicitly are encouraged to risk the public's health simply to gain an economic advantage over their law-abiding competitors. Even when a pesticide poisoning is diagnosed, the government's inquiry into the cause often takes far too long to complete, if it is finished at all. Delayed and incomplete investigations allow violators to escape detection, make the cause of the poisoning more difficult to find, and mean that some exposed people may never be identified. Appropriate penalties should be levied for pesticide-related violations that create health or environmental hazards, or pose a reasonable possibility of affecting health or the environment. Prompt and meaningful punishment for violators would remove the competitive advantage gained by businesses that currently cut corners in violation of the law, and would help prevent the illegal conduct that causes human health hazards in the first place.[56]

Pesticide Residues and Tolerances

A pesticide residue is the amount of pesticide on a food commodity after an application. A tolerance is the legal limit of pesticide residue allowed in or on a raw agricultural commodity and, in appropriate cases, on processed foods. The EPA sets tolerances to ensure pesticide residues are at safe levels. A tolerance is established through a process known as risk assessment. A tolerance must be established for any pesticide used on any crop. If the level of pesticide residues exceeds the tolerance, then the food is an illegal sale. The FDA and USDA are the agencies responsible for inspecting food and enforcing tolerances. If food is found to have pesticide residues exceeding an established tolerance, it is confiscated and destroyed.[57]

Food Residues

To what extent is our food contaminated with pesticide residues, and how much of a hazard is this? Long-lived pesticides such as DDT, other organochlorines, and parathion are most likely to leave persistent residues. Even though most persistent residues are banned in the United States and other wealthy countries, they are still present where previously used, are still used elsewhere, and show up on foods. Others, such as organophosphates, tend to break down so rapidly that they are unlikely to contaminate food unless applied to crops very close to harvest time. However, they are often more acutely toxic. Approved pesticides leave little residue on crops when used according to directions. National and international standards of acceptable residue levels are based on approved usage and maximum acceptable daily intake (MADI) levels; thus, theoretically, foods should be safe. However, it is not practical to monitor all crops and food shipments, residues are not always detected even when tests are done, and pesticides are not always used according to directions; thus, dangerous pesticide residues do make their way to the foods that people eat.[58]

Reasonable Certainty of No Harm

The Food Quality Protection Act of 1996 (FQPA) requires that tolerances be "safe," defined as "a reasonable certainty that no harm will result from aggregate exposure," including all exposure through diet and other non-occupational exposures, including drinking water, for which there is reliable information. It also distinguishes between cancer and non-cancer effects, consistent with EPA practice. The law establishes a single, health-based standard for all pesticide residues in all types of food, replacing the sometimes conflicting standards of the old law. There are no differences in the standards applicable to tolerances set for raw and processed food. Additional provisions ensure coordination with standards and actions under FIFRA for a more consistent regulatory scheme.[59]

Risk Assessment

Risk assessment is a process used by the EPA to determine if pesticide residues on food may prove harmful to human health. Toxicity and exposure are the two main components in risk assessment. Toxicity indicates the capacity of a pesticide to cause harm. Exposure describes how a pesticide will come in contact with the body and at what quantity and duration. A person can be exposed by eating, breathing, or touching pesticides. A pesticide can be very toxic, but exposure is necessary for there to be a health risk. The toxicity of a pesticide is usually determined by tests on laboratory animals. Scientists expose the animals to high levels of a pesticide to determine what health effects occur. The results of these studies give scientists the ability to determine the relative toxicity of pesticides in various species of animals. Determining accurate exposure to pesticides is a very difficult task. The exposure to a pesticide can be via multiple routes and for varying durations. Certain assumptions are made about the consumption of a food item under consideration. Often the exposure assessment is on the high end of the data range to account for the possibility that someone might consume a large quantity. In reality, people eat varying quantities of food and this variability is not accounted for in traditional risk assessment.[60]

The Delaney Clause

The Delaney Clause, named after Representative James J. Delaney, a New York Democrat, is a provision that prohibits without exception the use of any food additives in processed food that may cause cancer in humans. Before the 1996 FQPA, pesticides had been considered food additives and been subjected to the Delaney Clause. Although a well-intentioned provision, there were significant problems in applying the Delaney Clause to pesticide residues. If a pesticide that causes cancer in humans or laboratory animals is concentrated in ready-to-eat processed food at a level greater than the tolerance for the raw agricultural commodity, then the clause prohibited the setting of a tolerance. This had paradoxical effects in terms of food safety, since alternative pesticides could pose higher (non-cancer) risks, and the EPA allowed the same pesticide in other foods based on a determination that the risk was negligible. The Delaney Clause still applies to food additives, but under the FQPA pesticide residues are not considered food additives. Pesticide tolerances must be set to ensure reasonable certainty of no harm.[61]

Aggregate Exposure

Traditionally, the EPA has assessed human exposure to pesticides by individual chemicals and a single route of exposure. For instance, a person may be exposed to a pesticide through drinking water, eating food, and walking barefoot through recently sprayed grass. The EPA would examine each exposure route separately and report the risk separately, not combined. Under FQPA the EPA must consider all routes of exposure when setting food tolerances. To help visualize this concept, the term "risk cup" was coined to provide an analogy of total or aggregate pesticide exposure. Returning to the previous example, before FQPA there would have been three risk cups for each pesticide exposure through drinking water, eating food, and walking barefoot through recently sprayed grass. After FQPA, the EPA has only one risk cup that must account for all exposures to pesticides through water, food, and walking barefoot through recently sprayed grass. The risk cup is only so big and will allow only a finite amount of risk in the cup. When the risk cup becomes full, then any excess risk or exposure must be controlled. The size of the cup is determined by the definition of reasonable certainty of no harm. The EPA has developed an interim approach that assigns portions of the risk cup to specific pesticide exposure pathways. The risk cup is divided into 5 percent for residential exposure, 5 percent for outdoor exposure, 10 percent for drinking water exposure, and 80 percent for food exposure. These are the assumptions that will guide the EPA's tolerance setting decision-making process until other methods have been researched and developed.[62]

Additional Tenfold Safety Factor for Children

Prior to the passage of the FQPA, the EPA tolerance-setting process did not account for the special diet considerations of infants and children. Infants and children have different food consumption patterns and may detoxify pesticides they are exposed to differently. A National Academy of Science report recommended up to a tenfold safety factor be used in setting the tolerance for food to account for the special needs of children. For example: the current tolerance for a pesticide on apples is 100 parts per million (ppm). With the addition of a tenfold safety factor for children, the tolerance would now be set at 10 parts per million. The EPA has determined that there will not be an across-the-board tenfold safety factor added to every tolerance. The EPA will assess each tolerance and apply up to a tenfold safety factor on a case-by-case basis.[63]

The Common Mechanism of Toxicity

The common mechanism of toxicity describes how two or more pesticides produce the same adverse health effect. The FQPA requires the EPA to evaluate pesticide tolerances through a combined risk assessment for all pesticides that exhibit a common mechanism of toxicity (CMT). For example, malathion and diazinon are both organophosphate pesticides. If the EPA determines that malathion and diazinon have a common mechanism of toxicity, then any crops that use both pesticides will have to combine the risks of the two pesticides when setting tolerances. Historically, each pesticide would have been evaluated separately.[64]

Endocrine Disrupters

The endocrine system is a collection of glands that are located in several areas of the body. These glands release hormones into the bloodstream. The hormones travel to different locations in the body and act on specific "target" organs. If the endocrine system is disrupted, those organs will not receive the correct amount of hormones and might not function properly. Many think certain pesticides at low levels in the environment disrupt the endocrine system. The FQPA requires the EPA to develop a test to screen pesticides for potential endocrine disruption.[65]

Other Areas of Inquiry

In addition to topics already mentioned, this book will address a number of other issues involving pesticides, including the following inquiries: How will the risk of pesticide exposure for all Americans, but especially for children, be managed? What are the scientific and policy issues surrounding pesticide use and farmworker safety? How will the strategies embodied in Integrated Pest Management (IPM) reduce the health risks associated with pesticides? What accounts for the present inadequacy of pesticide safety measures? In what manner does pesticide toxicity and the hazards of "inert" ingredients exacerbate pesticide dangers? How are acute and chronic toxicity and associated uncertainties for humans assessed? Do the legal and regulatory frameworks under which the federal government establishes policies related to pesticide use produce effective results?

Future Prospects

The risks of acute poisoning and concerns about chronic impacts of exposure to pesticide residues in food continue to be debated. Natural resources can be degraded when pesticide residues in storm-water runoff enter streams or leach into groundwater. Pesticides that drift from the site of application to wildlife habitats may harm or kill non-target plants, birds, fish, or other wildlife. The mishandling of pesticides in storage facilities and in mixing and loading areas contributes to soil and water contamination.

Hundreds of years ago, when chemical pesticides were rare or nonexistent, farmers protected their crops through changes in growing practices and other actions. With their advent some six decades ago, pesticides became the primary means to control crop losses due to a variety of pests. In agriculture, pesticides will have to remain part of pest management strategies because alternatives or safer chemical pesticides may not be available or affordable for all farmers. Pesticides are likely to be an economic necessity for most farmers and growers. However, consumers continue to indicate support for a reduction in pesticide residues not only in food but also in the wider environment. Unfortunately, the new generation of environmentally friendly pesticides exhibits many of the same patterns as conventional pesticides, including resistance from targeted species and harm to non-targeted species. It is likely that the banning of traditional pesticides will continue and that those that remain will have further restrictions placed on how they can be used.

The EPA is approving new pesticides for use. However, these approvals often occur before scientists have the opportunity to determine how best to use them in IPM programs. While newer pesticides are typically safer to humans and the environment and are usually more selective, that is, they impact pests to a greater degree than natural enemies, they are also less effective than the products they replace.

The most obvious danger to human health from pesticides is through accidental poisonings. Chronic illness appears to arouse the greatest concern, especially the possibility of harm to children. What seems to worry people more is that long-term exposures to extremely small quantities of pesticides may be dangerous. Some experts argue that tiny amounts of pesticides in the foods people eat could lead to cancer and other illnesses that develop over a long period of exposure. Pesticides are in the middle of a tug-of-war. Because they are poisons, many people don't want them around, yet their value in protecting crops and combating pests cannot be denied.

Given this conflict, making precise predictions about pesticide use is difficult, but it is possible to discern some trends. Scientists and pesticide manufacturers will engage in a persistent quest to design and develop a range of safer pesticides, and evolve new and novel strategies to control insect pests and minimize crop damages. Clearly, the strong government regulation that took shape in the 1970s will continue. The United States is unlikely ever to go back to the days of government-encouraged spraying without limits, or allow pesticides to be invented and marketed without stringent requirements for testing and labeling. The rest of the world is moving in that direction as well. Change is a fact of life.

Notes

1. Lawrie Mott et al., "Our Children At Risk: The Five Worst Environmental Threats To Their Health, Chapter 5: Pesticides" (Washington, D.C.: Natural Resources Defense Council, September 16, 2004) (46): 141–146.

2. Robert Repetto and Sanjay S. Baluga, *Pesticides and the Immune System: The Public Health Risks* (Washington, D.C.: World Resources Institute, September 2004): 1,478–1,484.

3. Shanna H. Sean et al., "Pesticides and the Immune System," *Pesticide News* (June 1996): 15.

4. Jerome Blondell, "Epidemiology of Pesticide Poisonings in the United States with Special Reference to Occupational Cases," *Occupational Medicine: State of the Art Reviews* 12 (1997): 209–220.

5. T. L. Litovitz et al., "The 2000 Annual Report of the American Association of Poison Control Centers Toxic Exposure Surveillance System," *American Journal of Emergency Medicine* 19 (2001): 337–395.

6. S. H. Schuman and W. Simpson, "A Clinical Historical Overview of Pesticide Health Issues," *Occupational Medicine: State of the Art Reviews* 12 (1997): 203–207.

7. Lewis Regenstein, *America the Poisoned* (Washington, D.C.: Acropolis Books, 1982): 103.

8. League of Women Voters Education Fund, *America's Growing Dilemma: Pesticides in Food and Water* (Washington, D.C.: LWVEF, 1989): 1.

9. Richard Wiles, *Testimony Before the Subcommittee on Health and the Environment* (Washington, D.C:. Environmental Working Group, October 21, 1993).

10. *Pesticide Industry Sales and Usage, 1990–1991 Market Estimates* (Washington, D.C.: Office of Pesticide Programs, Economic Analysis Branch, Biological and Economic Analysis Division, EPA, Fall 1992): 2.

11. Ibid.

12. U.S. General Accounting Office, Report to the Chairman, Committee on Agriculture, Nutrition, and Forestry, United States Senate, *Pesticides: Limited Testing Finds Few Exported Unregistered Pesticide Violations on Imported Food* (Washington, D.C.: GAO, October 1993): 3.

13. David Pimental et al., "Environmental and Economic Cost of Pesticide Use," *BioScience* 42 (10) (1992): 750–760.

14. James R. Davis, Ross C. Brownson, and Richard Garcia, "Family Pesticide Use in the Home, Garden, Orchard, and Yard," *Archives of Environmental Contamination and Toxicology* 22 (1992): 260–266.

15. National Research Council, *Alternative Agriculture* (Washington, D.C.: National Academy Press, 1989): 123.

16. Ibid.

17. Ibid.: 121.

18. Ibid.: 123.

19. Public Broadcasting Corporation's *Frontline*, "In Our Children's Food," aired March 30, 2003, Martin Koughan, producer and director.

20. Nancy Blanpied, ed., *Farm Policy: The Politics of Soil, Surpluses, and Subsidies* (Washington, D.C.: Congressional Quarterly, Inc., 1984).

21. *Alternative Agriculture*. Op. cit. 123.

22. Ibid.: 125.

23. J. Buckley et al., "Pesticide Exposures in Children with Non-Hodgkins Lymphoma," *Cancer* 89 (11) (2000): 2,315–2,322.

24. S. Balk and K. Shea, eds., "A Partnership to Establish an Environmental Safety Net for Children," *Pediatrics* 112 (suppl.) (2003): 209–264.

25. M. S. Linet et al., "Interpreting Epidemiologic Research: Lessons from Studies of Childhood Cancer," *Pediatrics* 112 (suppl.) (2003): 218–232.

26. Paul Cotton, "Environmental Estrogenic Agents Area of Concern," *Journal of the American Medical Association* 27 (February 9, 1994): 414, 416.

27. Theo Colborn et al., "Developmental Effects of Endocrine-Disrupting Chemicals in Wildlife and Humans," *Environmental Health Perspectives* 101 (5) (October 1993): 319–384.

28. Janet Raloff, "The Gender Benders," *Science News* 145 (January 8, 1994): 24–27.

29. "Pesticide Usage—8.8 Pounds Per Capita." *Arkansas Pesticide News* 6 (October 1994): 1.

30. Nancy Blanpied, ed. Op. cit.

31. "The Poisons Around Us," *Synergy* (Summer 1998): 1.

32. T. Colborn et al. Op. cit.

33. C. Bernard and B. J. R. Philogene, "Insecticide Synergists—Role, Importance, and Perspectives," *Journal of Toxicology and Environmental Health* 38 (2) (February 2003): 199–223.

34. D. Pimental and H. Lehman, eds., *The Pesticide Question: Environment, Economics, and Ethics* (New York: Chapman & Hall, 1993).

35. C. K. Winter, "Pesticide Tolerances and Their Relevance as Safety Standards," *Regulatory Toxicology and Pharmacology* 15 (1992): 137–150.

36. S. G. Gilbert, lecture given at A Small Dose of Toxicology: How Chemicals Affect Your Health, a conference sponsored by the Northwest Center for Occupational Safety and Health, University of Washington, Seattle, October 17, 2001.

37. W. H. Hallenbeck and K. M. Cunningham-Burns, *Pesticide and Human Health* (New York: Springer-Verlag, 1985): 8–20.

38. Joel Grossman, "Dangers of Household Pesticides," *Environmental Health Perspectives* 103 (6) (June 1995): 550–554.

39. Northwest Coalition for Alternative to Pesticides, *Worst Kept Secrets: Toxic Inert Ingredients in Pesticides* (Seattle, WA: Northwest Coalition for Alternatives to Pesticides, 1998): 3.

40. Ken Toews, "Hidden Costs Mean Hidden Dangers," *Peace and Environmental News* (May 1992): 1.

41. Arnold L. Aspelin, *Pesticides Industry Sales and Usage: 1994 and 1995 Market Estimates* (Washington, D.C.: Environmental Protection Agency, Office of Pesticide Programs, 1997): 35.

42. David Pimental et al., "Environmental and Economic Cost." Op. cit.

43. Arnold L. Aspelin. Op. cit.

44. C. Cox, "Jimmy and Jane's Day: A Precautionary Tale," *Journal of Pesticide Reform* 18 (2) (1998).

45. GAO Report to the Ranking Minority Member, Committee on Government Affairs, U.S. Senate. *Pesticides Use, Effects, and Alternatives to Pesticides in Schools* (Washington, D.C.: GAO, November 1999).

46. W. F. Ritter, "Pesticide Contamination of Ground Water in the United States—A Review," *Journal of Environmental Science and Health* 25 (1990): 1–29.

47. D. W. Kolpin et al., "Occurrence of Pesticides in Shallow Groundwater of the United States: Initial Results from the National Water-Quality Assessment Program," *Environmental Science and Technology* 32 (5) (1998): 558–566.

48. W. B. Solley et al., *Estimated Use of Water in the United States in 1995* (Washington, D.C.: U.S. Geological Survey Circular 1200, 1998): 7.

49. B. Barles and J. Kotas, "Pesticides and the Nation's Ground Water," *EPA Journal* 13 (3) (1984): 43.

50. *Frontline*. Op. cit.

51. Lewis Regenstein. Op. cit.

52. Federal Insecticide, Fungicide, and Rodenticide Act (FIFRA) of 1947, as amended.

53. John E. Bonine and Thomas O. McGarity, *The Law of Environmental Protection* (St. Paul, MN: West Publishing Company, 1984).

54. F. Whitford et al., *Pesticides and the Law: A Guide to the Legal System* (West LaFayette, IN: Purdue University Cooperative Extension System, 1996): 14–16.

55. Leslie W. Touart and Anthony F. Maciorowski, "Information Needs For Pesticide Regulation in the United States," *Ecological Applications* 7 (4) (1997): 1,986–1,993.

56. T. H. Jukes, "People and Pesticides," *American Scientist* 51 (3) (1963): 355–361.

57. B. Eskanazi et al., "Exposures of Children to Organophosphate Pesticides and Their Potential Adverse Health Effects," *Environmental Health Perspectives* (1999) (Suppl. 3): 409–419.

58. J. H. Hotchkiss, "Pesticide Residue Controls to Ensure Food Safety," *Critical Reviews in Food Science and Nutrition* 15 (1992): 137–150.

59. T. H. Jukes. Op. cit.

60. P. S. Mead et al., "Food-Related Illness and Death in the United States," *Emerging Infectious Diseases* 5 (1999): 607–625.

61. T. H. Jukes, "The Delaney Clause: A 1990 Appraisal," *Priorities* (Winter 1991): 23–24.

62. C. Lowenherz et al., "Biological Monitoring of Organophosphates: Pesticide Exposure among Children of Agricultural Workers in Central Washington State," *Environmental Health Perspectives* 105 (12) (1997): 1,344–1,353.

63. A. G. Renwick, "Safety Factors and Establishment of Acceptable Daily Intakes," *Food Additives and Contaminants* 8 (2) (1991): 135–150.

64. M. Marinovich et al., "Effects of Pesticide Mixtures of In Vitro Nervous Cells: Comparison with Single Pesticides," *Toxicology* 108 (1996): 201–206.

65. T. Colborn et al. Op. cit.

Two

Pesticides in Agriculture

The Pesticide Dilemma

> Twenty-two years that I have been working in the fields, I've seen more ill-
> nesses, more children being born ill, more families that miss work because
> every day they have more problems, headaches. Sometimes their children are
> sick and they have to miss work. . . . We live in a depression. We don't know
> if it's because of the chemicals.
>
> —Laura Caballero[1]

The major source of workplace exposure to pesticides is in agriculture. The most
heavily exposed are workers who mix, load, apply, or otherwise handle the concentrated technical formulations. Farmworkers are exposed when cultivating and harvesting crops in fields, nurseries, and greenhouses, as well as transporting and handling
agricultural commodities in packing houses and storage facilities. Children living on
or near farms are exposed to disproportionately high amounts of dangerous pesticides, putting them at serious risk for adverse health effects. These children are likely
to have the highest exposure to pesticides of any group of people in the country.
Many of the children with the greatest pesticide exposure are from migrant farmworker families who are poor and usually people of color or recent immigrants.

There are approximately 17,000 pesticides on the market in the United States,
with about 800 in wide use. Pesticide use in this country alone amounts to 2.2 billion
pounds annually, or roughly 8.8 pounds per person. Virtually all of these pesticides
in use have undergone inadequate analysis of their safety. Most testing that has been
done has concentrated on acute toxicity and cancer-causing potential, ignoring possible endocrine-disrupting effects or damage to human immune systems.

Of the twenty-five most heavily used agricultural pesticides, five are toxic to the
nervous system; eighteen are skin, eye, or lung irritants; eleven have been classified by
the EPA as carcinogenic; seventeen cause genetic damage; and ten cause reproductive
problems in tests of laboratory animals. Annual use of pesticides causing each of these
types of health problems totals between 100 million and 400 million pounds.[2]

Total pesticide use, and the number of different chemicals applied, has increased substantially since the 1960s, when the first reliable records of pesticide use were established. Herbicide use has increased substantially and now accounts for approximately 75 percent of the total agricultural use of pesticides. Total insecticide usage has declined slightly, and a major shift in the types of compounds used has taken place, as organophosphorus and other insecticides have largely replaced organochlorine compounds. Fungicide use has increased slowly over the last two decades, and still represents only a small fraction—approximately 6 percent—of total agricultural pesticide use. Increased use of pesticides has resulted in increased crop production, lower maintenance costs, and control of public health hazards. However, concerns about the potential adverse effects of pesticides on the environment and human health also have grown.[3]

Overview of Farm Labor

Farm labor is seasonal and intensive. Planting, thinning, and harvesting are not year-round activities. However, they are crucial to crop production, and the time frame in which they must occur is determined by the seasons and the weather. Failure to perform any of these activities at the appropriate time can result in a lost crop. The urgency to accomplish tasks according to agricultural timetables compels farmworkers to labor in the fields in all seasons and in all weather conditions, including extreme heat, cold, rain, bright sun, and damp conditions.

Farmworkers' work hours accommodate the crops, not vice versa. Their work often requires stoop labor, working with the soil, climbing, carrying heavy loads, and making direct contact with plants. The crops and the soil are frequently treated with pesticides and chemical fertilizers. Some plants, such as tobacco and strawberries, exude chemicals that are toxic to humans or that can cause severe allergic reactions such as contact dermatitis.

There are anecdotal reports of farmworkers resorting to irrigation ditches and runoff ponds when safe water is not available for drinking and washing. Pesticides, chemical fertilizers, and organic wastes contaminate this water. Drinking and bathing in such water exposes farmworkers to potentially harmful chemicals, and also to waterborne pesticides.[4]

The estimated 4.2 million migrant and seasonal farmworkers in the United States constitute a population at risk for serious environmental and occupational illness and injury as well as health disparities typically associated with poverty. Although farmworkers are essential to the production of food in the United States, they have little power to control their work conditions. Farmworkers often make little more than minimum wage, seldom receive any employment benefits, and in many areas are not organized. Most farmworkers are immigrants and the national farmworker population has become increasingly Latino and Mexican during the past decade. In 1998, 81 percent of all migrant and seasonal farmworkers in the United States were foreign-born, and 95 percent of those were born in Mexico.

Although some areas of the United States (such as California and Florida) have routinely employed large numbers of Latino seasonal and migrant farmworkers, other areas have recently experienced a dramatic increase in those workers as family labor gives way to hired labor. In North Carolina, which ranks fifth in the size of its farmworker population, most farmworkers fifteen years ago were African American. Today only 10 percent are African American; most are Latino like the rest of the farmworker population in the United States. Pesticides are a major source of occupational injury and illness to which farmworkers are exposed.[5]

Young migrant and seasonal workers are the fastest-growing segment of the agricultural workforce. Many of them are entering an unfamiliar country and working in agriculture for the first time. In addition to their developmental needs for nutrition, rest, and education, young migrant and seasonal workers are totally dependent on adults for ensuring their health and safety while employed in agriculture.

There are nearly 400,000 young children in the United States who actually live on farms, and many of the additional five million agricultural workers living near farms have children. These people are extraordinarily diverse, ranging from family farmers to professional pesticide applicators to migrant farmworkers. Other groups of people who do not farm may also have pesticide exposure. For example, urban landscapers, pet groomers, and urban pesticide applicators share at least one important characteristic with farm families: they may bring pesticide residues home to their children.

Agricultural work is difficult and dangerous. Annual rates of work-related deaths among farmworkers are much greater than those for the general workforce. Migrant and seasonal farmworkers have exceptionally difficult working and living conditions and may suffer particularly high pesticide exposures. They bear the brunt of the risks and are most likely to be overlooked by scientists and regulators.[6] It is clear that the possibility for exposure to pesticides is greatest among farmworkers. While agricultural use of chemicals is restricted to a limited number of compounds, farming is one of the few industries in which chemicals are intentionally released into the environment because they kill things.

Occupational Safety and Health

Agricultural workers have an annual death rate that is five times greater than the national rate for all occupations combined. The magnitude of pesticide exposures and their impact on the health of farmworkers is unknown, particularly among ethnic minorities. Minorities are more likely to be subjected to adverse agricultural exposures than non-minorities. Assessments of acceptable exposure to pesticides cannot be the same as the acceptable daily intake of pesticides from dietary exposure since migrant farmworkers are much more likely to have heavier exposure to pesticides.

According to the USDA's own data, agriculture is one of the most accident-prone industries in the United States. Although the occupational fatality rate for all private-sector industries is 4.3 per 100,000 full-time employees, the rate for the broad category of agriculture, forestry, and fishing is 23.9. Other data sources indicate even higher accident and fatality rates in agriculture. The EPA estimates that 10,000 to 20,000 physician-diagnosed pesticide poisonings occur annually. Farmworkers, groundskeepers, pet groomers, fumigators, and other occupations are at risk for exposure to pesticides, including fungicides, herbicides, insecticides, rodenticides, and sensitizers. When it comes to undiagnosed illnesses, the EPA estimates that 300,000 farmworkers suffer acute pesticide poisoning each year. Anecdotal reports from clinicians indicate that many cases of pesticide poisoning are unreported because individuals do not seek treatment, or are misdiagnosed because the symptoms of pesticide poisoning can resemble those of viral infection. For descriptions of illnesses by occupation, industry, and pesticide-functional class, see Figures 2.1–2.3.

Pesticides must be registered for specific uses by the EPA. The agency considers the economic, social, and environmental risks and benefits of each pesticide before issuing the registration. Unfortunately, the pesticide manufacturers themselves provide the vast majority of the information the EPA uses to make its determinations, automatically creating a potential conflict of interest. Furthermore, data on older pesticides is considered incomplete by modern scientific standards, and the health effects of these substances are not fully understood.[7]

The OSHA Field Sanitation Standard

For most working people, it is taken for granted that sanitary facilities on the job, including operating toilets, potable drinking water, and hand-washing facilities, will

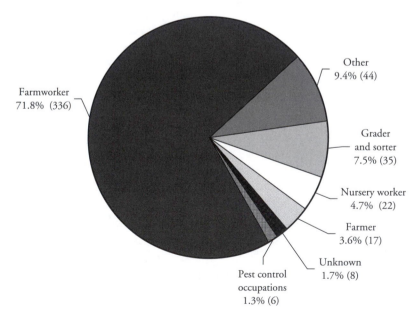

Figure 2.1 Distribution and Number of Pesticide-Related Illnesses among Agricultural Workers by Occupation. Sources: NIOSH 2002d; Calvert 2002.

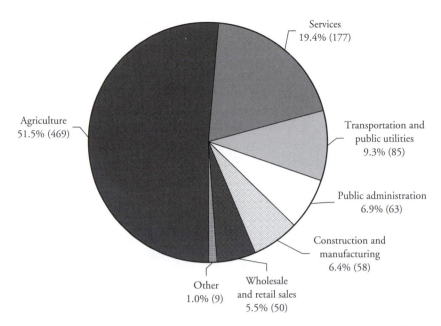

Figure 2.2 Distribution and Number of Pesticide-Related Illnesses by Industry, 1998–1999. Sources: NIOSH 2002d; Calvert 2002.

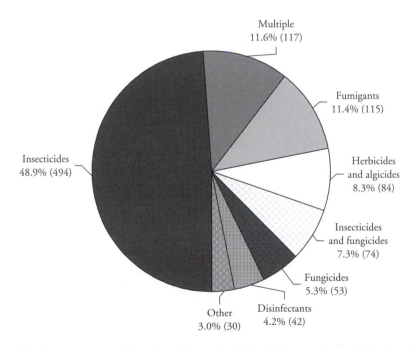

Figure 2.3 Distribution and Number of Pesticide-Related Illnesses by Pesticide Functional Class, 1998–1999. Sources: NIOSH 2002d; Calvert 2002.

be provided by employers. However, farmworkers are not accorded the same protections as other workers. The Occupational Safety and Health Administration (OSHA) issued a field sanitation standard in the late 1980s requiring toilets, drinking water, and hand-washing facilities. However, under federal law, employers of ten or fewer workers need not provide farmworkers with toilets, hand-washing facilities, or drinking water. The denial of such basic amenities is not just an affront to dignity but a serious public health issue. Women in particular are exposed to urinary and parasitic infections by the lack of these sanitary facilities. Plus, the denial of drinking water has resulted in preventable deaths in the fields from heatstroke. Sanitary facilities are automatically provided for in other occupations and should also be afforded farmworkers as a matter of federal law.[8]

For that matter, compliance with OSHA's standards and regulations has been poor. In 1990 OSHA found field sanitation violations in 60 percent of its field inspections. The fact that OSHA can afford to inspect only a small portion of establishments subject to the law raises questions as to the actual magnitude of noncompliance with its regulations. A North Carolina survey revealed that only 4 percent of farmworkers investigated had access to drinking water, hand-washing facilities, and toilets.

With regard to pesticides, both OSHA and the EPA have laws on the books that cover migrant and seasonal farmworkers. Because of possible jurisdictional difficulties, and due to overlap in the regulations, OSHA deferred its standard to the EPA's Worker Protection Standard. Although in 1983 the EPA determined that the Worker Protection Standard provided insufficient protection to farmworkers and was in need of revision, the revised standard scheduled to take effect in 1994 was postponed until 1995.[9]

The Worker Protection Standard

The EPA's Worker Protection Standard (WPS) is a regulation aimed at reducing the risk of pesticide poisonings and injuries among agricultural workers and pesticide handlers. The WPS contains requirements for labeling, pesticide safety training, notification of pesticide applications, use of personal protective equipment, restricted entry intervals following applications, posting and signs, decontamination supplies, and emergency medical assistance. Initially, the WPS was a very simple statement: workers were not allowed to enter the field until the sprays had dried or the dust had settled. The WPS was amended several times before it was finalized in 1995. This resulted in a very complex rule that is difficult for the agricultural community, both the farmers and the workers, to understand. It is very weak and poorly enforced. Most farmworkers have no idea what pesticide residues are on the crops they cultivate or harvest, or of their potential health effects.[10]

The WPS still leaves a significant number of workers unprotected. For example, the WPS does not adequately address the problem of drift of airborne pesticides onto adjacent fields where people may be working, or onto adjacent work camps where people may be living. Although the WPS requires that warning signs be posted, they may be posted immediately before a pesticide application, and required location of the signs is intended to target the workers on the farm being sprayed. However, there is no mechanism to ensure that workers in the field will be warned prior to the spraying of an adjacent field. If even a slight breeze is blowing in their direction, those workers will be subjected to potentially injurious exposures, despite the fact that all relevant laws have been obeyed. A survey of children working on farms in New York state revealed that nearly half had worked in fields still wet with pesticides, and over one-third had been sprayed directly or indirectly. Studies in Texas, Washington, and Florida indicated similar effects. In these states, 40 percent of all farmworkers had been sprayed directly or by pesticide drift. Further investigations demonstrate that fewer than 10 percent of farmworkers knew the symptoms of pesticide poisoning, understood the concept of pesticide entry intervals, or had received any training on how to protect themselves from pesticides. Therefore, it should come as no surprise that an estimated 80 percent of pesticide illnesses go unreported nationwide.[11]

Hopefully, the following comments of a Washington state fruit farmer are atypical, but if not, such violations undercut the intent of the WPS. He admits that occasionally he bends the rules on pesticide application and expects workers to reenter fields prematurely after spraying has taken place:

"The regulations are killing us. If we stuck to every rule there was, we could not do it," he says, frowning as he takes a long pull on a cigarette. "My workers, they go in and pick, and they're well aware of the pesticides we've sprayed; they know the damn stuff won't hurt 'em."[12]

No Protection for Children

Unfortunately, the current WPS does not consider pesticide exposures to children. No separate pesticide reentry intervals specifically for children have been established as yet even though recommendations have been submitted to the Department of Labor regarding minimum reentry times for ten- and eleven-year-olds working in fields of potatoes and strawberries. These intervals ranged from two to 120 days. They were adopted into regulations but ruled illegal by the U.S. Court of Appeals for the District of Columbia Circuit in 1980 in *National Association of Farmworkers Organizations v. Marshall*, 628 F.2d 604. Although children as young as ten can legally work in the fields, reentry intervals are calculated based on a theoretical 150-pound male. Specifically, the EPA uses a body weight of 154 pounds, except in the case of pesticides that have potential fetal developmental effects, in which case the body weight is 132 pounds to account for women of child-bearing years. There is no clear evidence that the REIs would protect farm-working children who weigh less than these amounts or are younger than twelve years of age.[13]

This is a deeply disturbing fact. The failure to make this issue the highest priority speaks volumes about the dedication, or rather the lack thereof, of the U.S. government to the health of working children. Moreover, the EPA has few guarantees that the protections afforded by the standards are actually being provided for farmworkers in general or to children who work in agriculture. Regional EPA offices have been inconsistent in setting goals for the number of work inspections that states should conduct, in defining what constitutes a worker protection inspection, and in overseeing and monitoring the states' implementation and enforcement of the standards.

Establishing Reentry Intervals

There are currently two approaches for setting reentry intervals. One might be considered the past approach and the other the future approach. In the first (or past) approach, the EPA's WPS established interim reentry intervals based only on acute toxicity, without any consideration of the crop, the work activity, or the degree of exposure. So if you think in terms of the simple equation that risk is equal to hazard times exposure (Risk = Hazard × Exposure), this approach takes into account only one-half of the equation. However, this approach is the basis for almost all reentry intervals currently in place in the United States.

In the second (or future) approach, the EPA's reregistration process requires the development of product, crop, and activity-specific reentry intervals based on the risk associated with any given use scenario. This approach takes into account all aspects of the equation that risk is equal to hazard times exposure (Risk = Hazard × Exposure). This is how reentry intervals are currently being set and will eventually be the basis for all reentry intervals.

The reentry intervals that are customarily seen on U.S. pesticide labels are set based on the requirements of the WPS. If a product has acute toxicity by the dermal route or due to eye or skin irritation, that places it in Toxicity Category I, which is a forty-eight-hour reentry interval. Toxicity Category II products receive a twenty-four-hour interval and Toxicity Categories III and IV receive twelve-hour reentry intervals.

The fallacy of this approach is that you may have a high exposure reentry activity involving a Toxicity Category III product that presents a greater risk than a low exposure reentry activity involving a Toxicity Category I product. Using the WPS approach, the activity with the higher risk in this case ends up with a shorter reentry interval.[14]

Safety Training Deficiencies

The daily dangers posed by pesticides are too often exacerbated by inadequate communication; agriculture is an industry in which management often does not speak the language of its workforce. Informing farmworkers by delivering uncomplicated safety information in clear, everyday language is often overlooked.

Few occupational groups are more in need of such training than farmworkers. In-house safety training on farms, when it does happen, is not always offered in a manner workers understand as spelled out in federal regulation. And because trainees are automatically considered to be in legal compliance at the end of the class, there is usually no test of actual comprehension. Federal and state regulations require farmworkers to receive pesticide safety training from their employers—whether growers or farmers. The training sessions, however, don't stress the law and are seen by environmentalists and public health advocates as inadequate.

The California Study

A study by the University of California, Davis Health and Safety Center was conducted to determine whether farmworkers are aware of new regulations mandating safeguards designed to protect them from illness or injury caused by occupational pesticide exposure. It also sought to determine whether and how they had received the required safety training, and whether they believed they were at risk of pesticide illness in their workplaces. Nearly 500 interviews were conducted in Spanish in two California counties in the summer of 1997. Fewer than one in five workers had ever heard of the WPS containing a basic "right to know," or even the EPA. Most of those who claimed to know something about either could not provide anything substantive upon closer questioning. Residents of two farm labor camps in Yolo County were the most likely to have received some training (66 percent), but in most cases it was provided by nonprofit agencies, not their employers. Only a relatively few farmworkers living at private camps had received training. Overall, only about 16 percent of farmworkers said that they had received on-the-job pesticide safety training. Clearly, these results suggest that news about the WPS and the "right to know" had not yet reached most farmworkers in California.[15]

Personal Protective Equipment Use

A series of National Cancer Institute–funded studies began in 1995 to assess pesticide-related safety practices of independent dairy farmers in Wisconsin and to identify ways to reduce their exposures. The first study surveyed a small focus group of farmers about both their perceptions of health risks and their use of protective gear such as gloves, goggles, and chemical-resistant aprons. The second investigation measured compliance with pesticide-specific protective gear requirements among 220 randomly selected dairy farmers.

Farmers were very aware of being exposed but fewer than 10 percent of them fully complied with the protective gear requirements the last time they had applied pesticides to their crops.

Seeking an educational solution, the investigators launched a third study in which data was collected on the exposures and safety practices of 400 farmers, and 100 randomly selected participants took part in a three-hour pesticide safety educational workshop. While those in the intervention group reported an increase of protective gear use and a decrease in number of pesticides used after six months, they reported no significant reduction in exposures or increase in full compliance with gear requirements. The researchers concluded that more intensive educational programs are necessary to achieve these goals.[16]

Poor Enforcement

Reentry intervals are intended to prevent farmers and farm labor contractors from sending harvest workers into fields for a specified number of hours after particular pesticides have been applied in order to permit the chemicals to degrade into less toxic substances. The field sanitation regulation requires farmers and contractors to provide drinking water and sanitation facilities, which can be utilized in cases of acute pesticide exposure. These safe work practices are woefully underenforced. For instance, less than half of the seventy high-profile California pesticides have reentry intervals of more than one day, and many have no reentry intervals at all. The protective equipment and sanitation requirements are widely ignored; a targeted enforcement effort documented the manner in which even the most elementary hygienic practices are disregarded. In California, less than 3 percent of all farms are inspected each year by the state, and in many other states, the inspections are even more rare. Without strong enforcement of existing standards, violations are likely to continue. The EPA should expeditiously reevaluate the WPS in order to determine whether it adequately protects the health of farmworkers. The EPA should, for example, consider using standardized data on sizes and age-specific weights and heights for modeling children's exposure when more specific information on children's exposures to individual pesticides may be lacking.[17]

The Fresno County Incident

In California, suspected pesticide-related illnesses and suspected work-related illnesses and injuries are reportable conditions. On July 31, 1998, the Occupational Health Branch of the California Department of Health Services (CDHS) received a report from the California Department of Pesticide Regulation (CDPR) of a pesticide exposure incident in Fresno County involving thirty-four farmworkers, both adults and minors. CDHS investigated this incident by reviewing medical records of the thirty-four workers and interviewing twenty-nine. The workers' ages ranged from thirteen to sixty-four years with a median age of thirty-one years. The findings indicated that the workers became ill after early reentry into a cotton field that had been sprayed with three pesticides. The primary pesticide used was carbofuran, which, when used on cotton, has a restricted entry interval (REI) of forty-eight hours and requires both posting of treated fields and oral notification of workers. Neither warning was provided. After weeding for approximately four hours, the workers were transported to a second field two and one-half miles away that had been sprayed two days earlier with three pesticides whose REI was twelve hours. Within approximately one-half hour of entering the second field, the workers began feeling ill and stopped working. The symptoms most commonly reported by the thirty-four workers were: nausea (97 percent), headache (94 percent), eye irritation (85 percent), muscle weakness (82 percent), tearing (68 percent), vomiting (79 percent), and salivation (56 percent).

Thirty workers were transported immediately to a medical clinic; the other four went home, showered, and sought medical care three to seventeen days later. All of the workers received hospital treatment for symptoms, and twenty-eight missed at least one day of work. The CDHS continued to monitor these workers to assess the acute and chronic effects associated with these pesticide overexposures. In this incident, workers entered a field at 6 A.M. to complete weeding begun the previous day. This was well below the required forty-eight-hour reentry interval and without labeling and oral notification. The results were moderately severe illness. The incident demonstrates that 1) posted and oral warnings based on the REI are necessary to prevent illness among workers performing hand labor in fields recently treated with pesticides, and 2) failure to adhere to an REI can result in serious health consequences for the exposed workers. No worker without prescribed protective clothing should enter a treated area to perform a hand-labor task until the REI expires. The length of the REI depends on the specific pesticide but generally can be no less than twelve hours. Additionally, this incident demonstrates that sole reliance on these control measures may be inadequate, creating a case for the substitution of safer, less-toxic alternative pesticides when feasible, or of integrated pest management techniques, where pesticide usage is prohibited.[18]

State Regulations

Most Federal Insecticide, Fungicide, and Rodenticide Act (FIFRA) activities have been delegated to the states. Many of the states have adopted the federal WPS, and others have adopted standards that are more restrictive than the federal standard. Standards adopted in Arizona and California are two examples of more restrictive regulations.

California's standard requires that employers of pesticide handlers have a written training program for their employees and that handlers be trained every year, rather than every five years. Cards used to certify EPA training are not accepted in California. Workers must be trained before they can enter treated fields. Employers must have a hazard communication program (including material safety data sheets) in place. They must also provide periodic cholinesterase blood tests under certain conditions. Employers must also provide a written respiratory protection program at the work site. They must require people working alone with pesticides labeled as dangerous to have contact with another person every two hours during daylight and every hour at night. Contractors must be notified about areas of pesticide applications and areas where entry is restricted (California Code of Regulations n.d., and the California Environmental Protection Agency n.d.).

Arizona's standard requires that contractors be notified of areas where pesticides are or are about to be applied, areas under restricted entry, and locations of the central posting of pesticide safety materials (Arizona 1998). Pesticide safety training includes instruction in how to file a complaint with the Arizona Department of Health Services.[19]

Farm operators and family members are exempt from most federal safety laws and regulations. However, raising the awareness of farm operators and family members to both the prohibited and the recommended safe practices is an important goal, and is one of the challenges for promoters of farm safety.

Non-Reporting of Inert Ingredients

The EPA requires that pesticide labels disclose only the product's active ingredients, that is, those toxic materials that kill the pest, weed, or other target organism. However, pesticides also contain many other ingredients, called "inert," which deliver the active ingredients to the target. Many of these may also be toxic, but the government does not require them to be identified on pesticide product labels. Moreover, states are preempted by the federal government from requiring such labeling for pesticides. In 1998, New York, Connecticut, Alaska, Massachusetts, and other states submitted a federal petition to the EPA to require full-product labeling of inert ingredients. Rather than responding to the petition, the EPA referred the matter to two advisory committees, neither of which has a definite timetable for resolving this pressing issue. There still have been no recommendations made to the EPA, and none are expected in the foreseeable future. This is an example of effective lobbying and political pressure by the agricultural chemical industry.[20]

Pesticides Become Hazardous Waste

Although "inert" ingredients are often more toxic than "active" ingredients, the public consumer and even professional applicators usually have no idea as to their contamination potential. Furthermore, it appears that some corporations may be using this avenue as a cheap form of hazardous waste disposal. This egregious policy came to light in 1991 when an EPA press officer told a reporter that allowing recycled hazardous waste in pesticides is "a way of disposing of hazardous materials." Two days later, when the reporter phoned to check the quotation, officials changed it to "a way of *using* [italics mine] hazardous materials." Either way, there can be no doubt that "recycling" hazardous waste into pesticides is a perfectly legal and EPA-approved way of "using" hazardous wastes.[21] A little-known exemption in the Resource Conservation and Recovery Act of 1976 (RCRA), the nation's basic hazardous waste law, allows hazardous wastes to be "recycled" into pesticides as "inert" ingredients.[22]

The Complete Circle

Not long after the above incident, it became evident that corporations were exploiting this legal loophole. GranTek, Inc., a Green Bay, Wisconsin, firm, takes contaminated sludges from Georgia Pacific Corporation's nearby paper mill. They dry the sludge, pelletize it, and send the pellets to a chemical company in Illinois, where they are used as the "inert" carrier ingredients for mosquito insecticides.

Ironically, the original sludge is created as part of a water pollution control process to prevent waste chemicals from entering an aquatic environment, Green Bay's Fox River. However, the GranTek pellets are spread back into wetlands and other aquatic environments in an effort to kill mosquitoes. This is a complete circle. The paper company sludge is contaminated with PCBs, dioxins, toxic metals, and a host of other toxic substances, but people who buy and use the mosquito insecticide will never know this. GranTek pellets are also used for kitty litter and as a carrier for livestock pharmaceuticals. Those customers are also uninformed.[23]

This is vivid proof that it becomes impossible for pesticide users, whether they are government agencies, businesses, or homeowners, to accurately understand the hazards of a pesticide product they are proposing to use if they don't know its ingredients. "Inert" ingredients also pose a crucial ethical issue. We are all exposed to pesticides on a daily basis whether we like it or not. Given this situation, the very least that can be done is to ensure that we have complete and easily and publicly accessible information about all of the ingredients in pesticide products.

The So-Called Right to Know

Historically, farmworkers have been excluded from the right to know the names of the chemicals they work with or from training on how to protect themselves. Thus, farmworkers were the only occupational group excluded from the OSHA's Hazard Communication Standard (HCS). The WPS, as has been noted, requires safety training for all workers who enter crop fields where pesticides have been applied, and explicitly grants certain rights to workers, including a basic "right to know." Also, under state law, growers and farm labor contractors are required to inform workers of the risks they face and to train them in safe handling techniques. Written illness prevention plans are formally required. These "right-to-know" provisions are supported, in principle, by the "right-to-act" provisions of federal and state labor law, which guarantees to workers the right to join labor unions and bargain collectively with employers. However, the "right-to-know" movement among industrial workers and urban communities exposed to toxic chemicals has exerted a modest but beneficial impact on public policy toward pesticide-exposed farmworkers. Yet, these worker-oriented regulations have not always been observed in practice.[24] The reality is that the vast majority of farmworkers have no idea what chemicals are used where they work, much less what the specific health effects are. The present challenge is to develop ways to train farmworkers effectively so they can recognize the dangers of the pesticides they come into contact with, take measures to protect themselves, exercise their rights under the law, and work together to resolve problems when they are identified.

Primary Care Providers and Pesticide Issues

What is the knowledge and awareness of pesticide issues in the educational and practice settings of primary care providers? A primary care provider is defined as physician, nurse, nurse practitioner, physician assistant, nurse-midwife, or community health worker specializing in one of the following areas: family medicine, internal medicine, pediatrics, obstetrics/gynecology, emergency medicine, or public health. Americans look to their primary care providers for guidance on health concerns.

Public concern about pesticides can come from a variety of sources. Patients may have heard about pesticide issues in the national or local news, or read about the health effects of acute or chronic exposure to pesticides. Concerned patients may turn to their primary care provider for answers about their own risks of illnesses from exposure to pesticides. They may question providers about acute health effects as well as potential chronic effects, such as cancer, birth defects, reproductive effects, or other conditions resulting from pesticide exposure. Primary care providers need to be prepared to recognize, manage, and prevent pesticide-related health conditions in their patients and communities. By helping patients recognize symptoms of pesticide-related illnesses, and by problem-solving and evaluating risks from pesticides, primary care providers can help patients reduce exposure and prevent future exposures.

While some progress has been made in introducing environmental health issues into the curriculum of medical and nursing schools, most health professionals still do not have adequate knowledge and tools to address patient and community concerns. A survey of environmental medicine content in U.S. medical schools found that 75 percent of medical schools require only about seven hours of study in environmental medicine over four years of education. Consequently, once in practice, physicians' awareness and understanding of pesticide-related illnesses may or may not ever increase.[25]

The current lack of adequate environmental education for health care providers sounds an alarm for leaders in the health care community, whose professionals are not prepared to deal with pesticide-related illnesses as they are presented. Primary care providers are on the frontline of health care and therefore can play a key role in identifying, treating, and preventing potential pesticide exposure and poisonings.

Importance of Exposure Histories

The clinical history is an essential part of data collection and doctor-patient communication. The environmental history, including questions eliciting concerns and probing environmental hazards to which a patient is exposed, should be included in the routine medical history. However, physicians who do ask about environmental exposures usually limit their inquiries to lead and environmental tobacco smoke.[26]

In some situations where exposures are complex or multiple and/or symptoms atypical, it is important to consider consultation with clinical toxicologists or specialists in environmental and occupational medicine. Local poison control centers should also be consulted when there are questions about diagnosis and treatment.[27]

Although the clinician's concerns deal primarily with pesticide-related diseases and injury, the approach to identifying exposures is similar regardless of the specific hazard involved. It is important to ascertain whether other non-pesticide exposures are involved because of potential interactions between these hazards and the pesticide of interest (e.g., pesticide intoxication and heat stress in agricultural field workers). Few primary care providers ask patients the questions that would be likely to alert them to the possibility of pesticide-related illnesses. Although it is important for primary care providers to take environmental histories, a full environmental history can sometimes take up an entire patient visit. However, getting primary care providers to ask just a few simple questions, such as "Where do you work?" and "Do you think your problems are related to something that happened at work?" could go a long way toward answering pesticide-related health concerns about low-dose chronic effects as well as acute high-dose poisonings and effects on children.[28] Therefore, it is essential to obtain an adequate history of any environmental or occupational exposure which could cause disease or exacerbate an existing medical condition.

In many instances, rural health care providers possess neither the knowledge nor the training to record a proper medical history of a person's work exposure or the incident that led to the injury or illness. Furthermore, migrant clinicians either do not have access to prevalence data for specific kinds of injuries and illnesses or are unaware that such data exists. In other cases, either time constraints or an employer's unwillingness to cooperate prevent a physician from learning more about the origin of an individual's particular health problem. As a result, the migrant health clinic usually sees the hired farmworker on only the occasion of illness or exposure, and the clinician loses any opportunity to examine the long-term effects of a given injury or illness.[29]

Cultural origin is another obstacle blocking successful intervention by clinics and rural health facilities. Primarily due to language and cultural differences, farmworkers and clinicians may have trouble communicating with each other. Differences in terminology can affect a clinician's ability to take an accurate health history from a farmworker. In addition, many hired farmworkers hold biases against Western medicine, and as a result do not readily accept the advice of health care providers.[30]

Data Limitations

As is the case with the reporting of occupational injuries and fatalities in non-farming activities, there are serious data-gathering deficiencies in the reporting of pesticide-related illnesses and deaths. In the absence of comprehensive national information, the EPA uses four databases to provide some indication of the extent of acute pesticide incidents and illnesses. These databases are: 1) the American Association of Poison Control Centers' Toxic Exposure Surveillance System, 2) the data reported to the EPA under FIFRA, 3) the National Pesticide Telecommunications Network, a cooperative effort between the EPA and Oregon State University, and 4) the California Pesticide Illness Surveillance Program. However, each of these databases has its limitations:

- The American Association of Poison Control Centers maintains information on poison exposures. However, its database does not isolate pesticide exposures that occurred in agricultural work (or from any other occupation). In addition, some poison control centers do not report to the national database, and reports that poison control centers receive by telephone may lack medical confirmation.

- Under section 6(a)(2) of FIFRA, registrants are required to submit information they obtain about unreasonable adverse effects of their pesticide products. The 6(a)(2) database was designed to gather information on the effects of pesticides rather than on the extent of pesticide incidents. Therefore, the database contains detailed reports on serious and rare incidents, but little information on less-serious incidents.

- The National Pesticide Telecommunications Network is a toll-free telephone service that provides the general public and health professionals with information on pesticide health and safety and pesticide incidents. While the network categorizes pesticides by the age, sex, and occupation of the affected person, the network's data rely on self-reporting, and most of the information has not been verified or substantiated by independent investigation, laboratory analysis, or any other means. Moreover, many farmworkers, particularly migrant or seasonal workers, may not have ready access to a telephone to report pesticide incidents.

- The California Pesticide Illness Surveillance Program, often cited as the most comprehensive state reporting system, obtains most of its case reports through the state's workers' compensation system. Therefore, illnesses that occur in farm children, who are not officially workers, are unlikely to be reported in this system. Also, according to the EPA and farmworker advocacy groups, farmworkers may be reluctant to report pesticide exposures because of the potential for retaliatory actions such as the loss of jobs or pay cuts.

Notwithstanding the limitations of California's program, the EPA used this information in 1999 to make a nationwide estimate that there were 10,000 to 20,000 incidents of physician-diagnosed illnesses and injuries per year in farm work. However, the EPA recognized that its estimate represented serious underreporting (other estimates are as high as 300,000, as previously mentioned). Moreover, according to officials from the California Department of Pesticide Regulation, because California's crops and pesticide regulations are different from those of other states, it is inappropriate to extrapolate California's data to the rest of the nation. In addition, there are other reasons why acute pesticide incidents are underreported, including farmworkers' hesitancy to seek medical care for financial reasons and physicians' misdiagnoses or failure to report incidents.[31]

Quality Problems: Enforcement and Compliance Data

Good information is fundamental to effective management and public confidence in government agencies. For the last three decades, however, the EPA and most state environmental agencies have relied on data about enforcement activities that do not actually reveal how well the environment is doing, or how well the regulated community is obeying environmental laws. To the extent that these data measure enforcement or other governmental performance, they are much more likely to be misleading than useful.

The data on which the EPA and state agencies currently rely relate almost exclusively to activities: the number of permits issued, inspections conducted, enforcement actions initiated, and penalty dollars collected. For many years, these data have served as the basis for management decisions and oversight of agency performance. When the EPA delegated responsibility for implementing national environmental programs to the states and provided them with funding to do so, it created data systems to track these activities, and it used activities data to hold states accountable for proper use of federal funds and implementation of federal laws, including timely and appropriate enforcement.

Using these same data, the EPA's inspector general and an environmental advocacy group have recently concluded that state enforcement activities have declined, raising concerns that states have de-emphasized enforcement to the detriment of the environment. However, the Environmental Council of the States (ECOS), a national non-profit, non-partisan association of state environmental agency leaders, challenged these conclusions. It argued that state data in the EPA systems were often incomplete and inaccurate, did not reflect the full range of state compliance activities, and revealed little about whether environmental conditions are improving.[32]

Complications Stemming from Misdiagnoses

Farmers and farmworkers suffer from pesticide exposure and other ailments that often are misdiagnosed or improperly treated because doctors never have faced the malady or the patients don't think to pass along relevant information. Physicians, for instance, commonly misdiagnose as pneumonia lung irritations or infections that actually result from pesticides or other toxic substances. Or they see the farmworker before symptoms begin and send him or her home before fluid buildup and breathing difficulties start. Primary care professionals who can recognize a potential pesticide exposure are more likely to make the correct diagnoses. They need to be familiar with the settings that predispose patients to pesticide exposure, the symptoms associated with these exposures, and appropriate diagnostic methods. Yet many health professionals receive little training in pesticide health issues. The following cases illustrate that further illnesses could have been avoided with properly educated and trained health care providers.

The EPA regulates an organophosphate insecticide called methyl parathion for use on specific crops. During the 1980s and 1990s, methyl parathion was used widely by unlicensed applicators. One published report describes methyl parathion–related illness among several siblings, two of whom died. Approximately two days before these children were correctly diagnosed, five of them were seen by their local physician and sent back to their contaminated homes with a mistaken diagnosis of viral gastroenteritis.[33]

In another case of organophosphate poisoning, a group of thirty infants and children was poisoned by organophosphates and carbamates. Of twenty cases transferred to Children's Medical Center in Dallas, sixteen (80 percent) had an incorrect transfer diagnosis ranging from encephalopathy to seizure disorders to pneumonia to whooping cough.[34]

There must be some specialized training for doctors who work in rural medicine. South Carolina may have the best system in the nation, the South Carolina Rural Health Research Center. A network of experts helps physicians by teaching them to spot peculiar agricultural illnesses or by providing information for treatment. Under this network, physicians are linked with various extension agents, poison specialists, and other experts who have experience treating specific problems. To make a real difference, farmers and farmworkers must also learn what they need to tell doctors. When a farmworker goes to a doctor for treatment, it is up to that individual to say, "I work on a farm." Then it's up to the doctor to ask the questions. A farmworker who thinks he or she has problems because of exposure to some sort of toxic substance also needs to learn to urge the physician to contact poison control centers because those institutions often have suggestions for treatments that are immediately available.[35]

Financial Problems

The EPA has concluded that using existing surveys, particularly the Consumer Product Safety Commission's (CPSC) National Electronic Injury Surveillance System (NEISS) and the National Center for Health Statistics' National Hospital Discharge Survey (NHDS), and supplementing them with additional data collection specific to pesticides, as well as increasing coverage of hospitals in rural areas, would be more cost effective than initiating a new data collection system. However, the EPA never allocated funding to expand data collection and coverage of hospitals in rural areas, and the agency has not collected hospital emergency room data since 1987.[36]

With respect to the migrant health care centers, a lack of sufficient funding has hampered their ability to effectively serve farmworkers. Financial burdens due to cutbacks in migrant clinic funding have prevented many clinics from improving their health care services or expanding their knowledge about these types of patients. Without sufficient funding, many migrant health clinics cannot invest in the laboratory equipment necessary to make correct evaluations of work-related illnesses and injuries such as pesticide exposure. In addition, many large companies maintain contracts with private physicians and refer their employees who become injured or ill while working to them. This removal of a significant number of farmworkers from the patient pool treated by migrant health clinics creates further economic difficulties for clinicians. For the few farmworkers who remain, the significant amount of paperwork involved with workers' compensation claims, coupled with low reimbursement and the fear of litigation, may deter them from accepting such cases.

Also, funding for health surveillance projects has on occasion been inadequate. Without the necessary dollars, it is impossible for health projects to enlarge their scope and cover greater portions of the farmworker population. Likewise, insufficient funding prevents the development of new data collection systems that use more active surveillance techniques to document problems that are rarely seen at clinical facilities.[37]

Monitoring Pesticide Exposure

Workers who apply and mix pesticides are at special risk of systemic pesticide illness. Both acute and chronic exposure can occur from spillage or by environmental contamination of clothing. One serious health problem develops when cholinesterase levels in the body drop to low levels after pesticide exposure. The resulting nervous system malfunction produces pesticide-poisoning symptoms such as fatigue, light-headedness, nausea, vomiting, headaches, and seizures. If levels decrease too much, subsequent exposure to organophosphate or carbamate insecticides can result in death.

Cholinesterase is an enzyme essential for normal functioning of the nervous system. It inactivates the chemical messenger acetylcholine, which is normally active at the junctions between nerves and muscles, between many nerves and glands, and at the synapses or connections between certain nerves in the central nervous system.

Biological monitoring is the means by which absorption of a pesticide is proven. This is in contrast to environmental or atmospheric monitoring, which reveals the level of external contamination. In general, biological monitoring uses measurement of pesticide levels in blood, urine, saliva, breath, or hair as an indication of the amount of pesticide or other chemical that has been absorbed by all the routes of exposure, such as inhalation, ingestion, or through the skin. Nevertheless, it should be stressed that to evaluate workplace conditions on a continuous basis, emphasis should be placed on environmental monitoring complemented by biological monitoring. The latter has been found to be valuable in assessing the effectiveness of protective clothing and respirators as well as a measure of worker compliance with safety procedures.[38]

A basic monitoring system would periodically test cholinesterase levels in the blood of those at risk for cumulative exposure and insecticide poisoning. Blood samples can be drawn at a clinic and sent to a laboratory for evaluation or the entire procedure can be performed at the work site using field test kits. Workers shown to have dangerously low levels are then identified and reassigned to prevent further exposures until their depressed cholinesterase levels rise closer to normal levels.[39]

The Washington State Experience

Medical Monitoring: Documenting Toxic Exposures and Their Consequences.

After nearly twenty years of struggle and a state supreme court victory, farmworkers in Washington state who regularly handle organophosphate (OP) and carbamate (CB) pesticides finally received medical monitoring in 2004. Blood tests were taken before the spray season to establish each worker's normal levels of cholinesterase, which is lowered by OPs and CBs. Follow-up blood tests were conducted during spray season. When cholinesterase levels declined by more than 20 percent, employers were required to do workplace safety audits to identify causes of exposure. When levels declined by 30 percent or more for one type of test (red blood cell) or 40 percent or more for a different test (plasma serum), employers were required to remove workers from handling these pesticides and doing other tasks with high exposures.

First-Year Results: One in Five Workers with Significant Nervous System Impacts. Over the course of the spray season, 123 (20.6 percent) pesticide handlers out of 580 who received both baseline and follow-up tests had depressions in their cholinesterase levels of more than 20 percent. Of these, twenty-six (4.4 percent of the 580 workers) had depressions low enough to trigger removal. Depression rates were even higher early in the spray season, when one in four workers had action-level depressions and more than 6 percent needed to be removed. Serious depressions were likely undercounted because: 1) according to scientists who reviewed the program, there is a high risk of "false negatives" (test results failing to identify actual significantly lower levels of cholinesterase); 2) most baseline tests were run long after blood samples were taken—cholinesterase levels in these samples likely declined before the tests were run; and 3) some workers reportedly declined monitoring due to fear of retaliation by employers.

The Department of Labor and Industry's Inadequate Response. A major purpose of monitoring is to ensure swift audits and removals to prevent further exposures and injuries. Nonetheless, the Washington state Department of Labor and Industry decided to offer consultations to employers rather than to exercise its enforcement authority. This resulted in long delays between when agency consultation staff learned of depressions in cholinesterase levels and when workplace visits took place. The average time between the cholinesterase test results and inspections of workplaces was 34.5 days for workplaces requiring audits and thirty-five days for those where workers had to be removed. Often seven or more weeks had already elapsed.

Although some advocacy groups viewed the state experience as a clear indication that the use of the most neurotoxic pesticides should be phased out, others saw it as proof that protective measures are keeping farmworkers safe. Farmworkers play a vital role in Washington's agriculture. Results of the new medical monitoring program and recent studies reveal the steep price these workers and their families pay as the result of the industry's reliance on highly toxic pesticides. Farmworker protection advocates hope that the results of the state's first official biomonitoring study will help convince legislators to phase out the use of the most neurotoxic agricultural pesticides—with limited exceptions—by 2012.[40]

Pesticide Toxicity

For all pesticides to be effective against the pests they are intended to control, they must be biologically active, or toxic. Because pesticides are toxic, they are also potentially hazardous to humans and animals. For example, any pesticide can be poisonous or toxic if absorbed in excessive amounts. Pesticides can cause skin or eye damage (topical effects) and also induce allergic responses. However, if used according to labelled directions and with the proper personal protective equipment (PPE), pesticides can be used safely. For this reason, people who use pesticides or regularly come in contact with them must understand the relative toxicity and the potential health effects of the products they use. The risk of exposure to pesticides can be illustrated with the following simple equation:

Hazard of Pesticide Use = Toxicity × Actual Exposure

Toxicity is a measure of the ability of a pesticide to cause injury, which is a property of the chemical itself. Pesticide toxicity is determined by exposing test animals (usually rats, mice, rabbits, and dogs) to different dosages of the active ingredient. Tests are also done with each different formulation of the product, for example, liquids, dusts, and granulars. Pesticide toxicities are listed in milligrams of exposure to kilograms of animal body weight. By understanding the difference in toxicity levels of pesticides, a user can minimize potential hazards by selecting the pesticide with the lowest toxicity that still controls the targeted pest.

Applicators may have little or no control over the availability of low-toxicity products or the toxicity of specifically formulated products. However, exposure can be significantly reduced or nearly eliminated by using personal protective clothing and equipment. For example, more than 90 percent of all pesticide exposure comes from dermal exposure, primarily to the hands and forearms. Wearing chemically resistant gloves can reduce this exposure by at least 90 percent. Therefore, an applicator can lower a pesticide's hazards to an insignificant level by using the correct PPE.

Signal Words

Acute toxicities are the basis for assigning pesticides to a toxicity category and selecting the appropriate signal word for the product label. Pesticides that are classified as "highly toxic" on the basis of oral, dermal, or inhalation toxicity must have the signal words DANGER and POISON (in large red letters) and a graphic of a skull and crossbones prominently displayed on their package labels. PELIGRO, the Spanish word for danger, must also appear on the labels of highly toxic chemicals. Acute oral LD_{50} values for pesticide products in this group range from a trace amount to 50 mg/kg of body weight. An exposure to only a few drops of a highly toxic material taken orally could be fatal to a 150-pound person. Some pesticide products are labeled with the signal word DANGER without a skull and crossbones. A DANGER label does not provide information about the chemical's LD_{50} value. Instead, this signal word alerts users of potentially more severe skin or eye effects caused by the product's irritant or corrosive properties. Pesticide products considered "moderately toxic" must have the signal words WARNING and AVISO (the Spanish word for notice or warning) displayed on their labels. Acute oral LD_{50} values range from 50 to 500 mg/kg. Exposure to just one teaspoon to one ounce could be fatal to a 150-pound person. Pesticide products classified as either "slightly toxic" or "relatively nontoxic" are required to have the signal word CAUTION (PRUDENCIA) on their labels. Acute oral LD_{50} values are greater than 500 mg/kg.[41]

Pesticide Synergy: The Whole Is Greater Than the Sum of the Parts (or When 1 + 1 = 10)

A study published in *Science* showed that pesticides may be thousands of times more potent than previously thought. It demonstrated that, when tested alone, each of two particular organochlorine pesticides had to be at concentrations on the order of 100,000 times greater than natural estrogen to cause responses in yeast cells that reacted to estrogen. Yet the same two organochlorines mixed together required concentrations of only between ten and 100 times more than natural estrogen to induce the same response. Thus, exposure to multiple pesticides may be thousands of times more potent in mimicking estrogen than was previously thought. Other research found that 30 percent of apples contain at least three pesticide residues. These findings sent a chill through the EPA, which suddenly faced the possibility that all their safety tests of single chemicals were suspect.

Considering the diversity of pesticides found in our environment, the increased potency of combined pesticides raises many questions. Does this mean that current tolerance levels set for individual pesticide residues are actually far above dangerous limits when combined? What are the implications for the unborn, young children, and adolescents? In light of these findings, how should pesticide use be assessed? How should new chemicals be evaluated? In addition, what are the implications for future pesticide policy formulation?[42]

It appears, therefore, that synergistic effects between multiple pesticides and/or other chemicals represent one of the greatest gaps in the EPA's ability to protect the public from adverse health effects associated with pesticide use and exposure. The government recognizes that pesticide exposures occur in combinations and as unique events, yet its rules and regulations test only a limited number of possible interactions. Given that there are more than 875 active ingredients currently registered for use, it would be impossible to examine all possible combinations, but we must start somewhere. One approach would be to prioritize analyzing pesticides most likely to act in combination. This would include testing groups of pesticides that are frequently used on the same crops, such as atrazine and chlorpyrifos, which are among the most common herbicides and insecticides applied to corn.[43]

Pest Resistance

Genetic variations within pest populations leave some naturally resistant to pesticides. If pests have not been previously exposed to a new pesticide, most individual pests in the population are susceptible, but some individuals are resistant. Pesticides used to control the pest will kill most susceptible individuals, but the few resistant pests will survive. As a result, the proportion of resistant individuals in the population increases. Repeated selection of resistant pests will ensure that every succeeding generation will have a higher proportion of resistant individuals than the original population. Eventually, after repeated and more intensive use of the same pesticide to the same pest population, the pesticide becomes ineffective. Unfortunately, even under ideal conditions, pests can become resistant to pesticides.

Cross-Resistance

Cross-resistance may occur where a pest develops resistance to two or more pesticides after exposure to just one. For example, resistance to dieldrin confers cross-resistance to other related compounds such as chlordane and heptachlor.

In the last decade, the number of weed species known to be resistant to herbicides rose from forty-eight to 270, and the number of plant diseases resistant to fungicides grew from 100 to 150. Resistance to insecticides is so common—more than 500 species—that nobody is really keeping score. Unfortunately, pesticides often kill off pests' natural enemies along with them. With their natural enemies eliminated, there is little to prevent recovered pest populations from exploding to larger, more damaging numbers than existed before pesticides were applied. Additional chemical pesticide treatments only repeat this cycle.

Secondary Pests

Some potential pests that are normally kept under firm control by their natural enemies become actual pests after their natural enemies are killed by pesticides. Mite outbreaks after pesticide applications are a classic example.

Adding to this scenario the intractable problem of pesticide resistance by insects and disease organisms, it becomes clear why so much attention is being paid to the development and adoption of pest management approaches aimed at reduction in pesticide use. There is one basic rule with pesticides in relation to resistance: *avoid unnecessary use*. If resistance does occur the simplest procedure is to use an alternative registered pesticide to which there is no cross-resistance. Proven strategies either to prevent or delay the development of resistance do not exist at present, either for plant disease agents or insect pests. Theoretically, the rotation of different chemical groups should delay resistance, but long-term experimentation under practical conditions is almost impossible to carry out. The problems of implementation are formidable.[44] As of 1999, pest resistance to pesticides was estimated to cost U.S. agriculture about $1.5 billion per year in increased pesticide costs and decreased crop yields. On average developing a new pesticide costs $80 million, while a pest typically develops resistance in only ten to twenty-five years, after which the pesticide's utility decreases.

Continued dependence on chemicals as the sole method of pest control is a sure recipe for the ongoing development of pest resistance.[45]

Underreporting of Illnesses and Injuries

Mild signs of acute pesticide poisoning, such as nausea, vomiting, diarrhea, or wheezing, are often not recognized as being potentially linked to pesticide toxicity. Rashes and other skin reactions are other major manifestations of pesticide toxicity that are often misdiagnosed. The American Association of Poison Control Centers reported 97,278 calls about pesticide poisonings in 1996. Half of the reported poisonings involved children under six years of age. Occupational pesticide poisonings are required to be reported in California, and there are approximately 1,500 reported cases per year. Annually, national occupational pesticide poisonings among agricultural workers have been estimated at anywhere from 10,000 to 40,000 physician-diagnosed pesticide illnesses and injuries. However, these statistics do not include the children of agricultural workers.

Research has shown that current estimates based on occupational surveillance or poison control centers may greatly underestimate the number of pesticide poisonings. Follow-up evaluations of poisoned workers in California discovered that 40 percent of exposure incidents also involved co-workers who did not seek medical treatment for various reasons, suggesting that the total burden of illness is grossly underreported. Poison control centers are commonly called after accidental ingestions or spills of pesticides in the home, but are less frequently called when illnesses occur after routine agricultural pesticide exposure.[46]

The following are some examples of potentially harmful pesticides, along with their possible side effects:

- Atrazine, a weed killer sprayed on crops such as corn, has been reported by researchers to cause sexual abnormalities in frogs. Another study found that the herbicide caused high rates of prostate cancer among workers at a plant that produces it. It may also disrupt the ordinary production of human hormones.

- 2,4-D, claimed to be one of the safest pesticides on the market, has been reported to increase the risk of a number of health problems, including cancer, fetal and birth defects, liver and kidney injury, leukemia, and tumors.

- Paraquat, a commonly used herbicide, can damage the respiratory, circulatory, or muscular systems, as well as the brain and the heart.[47]

Health Impacts Among Farm Children

Much of the evidence of the chronic effects of pesticide exposure is based on studies of adult workers who are exposed to a mixture of chemicals every day, making it difficult to pinpoint their exposure to specific pesticides. Little research has been done directly on children, and even less on farm children. Nearly all of the epidemiological studies on children's health and pesticide exposures were done on the general, non-farming population. These studies would likely underestimate the health impacts that would be expected for highly exposed subpopulations of young people, especially farm children. Some studies did look at the children of parents who work in jobs that may involve pesticide exposure; however, the child's exposure was almost never directly assessed, but instead was indirectly estimated based only on the parent's job title. Such a technique is likely to lead to misclassification of exposures and underestimation of health effects. Thus, health impacts among farm children are likely much greater than those described in most of the scientific research to date. Because of the health effects of pesticides, it is important to identify the sources and levels of exposure to these chemicals in order to protect the most highly exposed children from these dangerous substances.

Much remains unknown about the risks faced by children in agriculture, and enforcement of pesticide protection standards for farmworkers is patchy and unsystematic. Children are known to be more vulnerable to the effects of pesticides, but there is a lack of data regarding children's exposures and the precise effects of pesticides on their health.[48]

Perceptions of Growers and Extension Agents Toward Farmworker Pesticide Exposure

Growers have more control over their own pesticide exposure risk than do the farmworkers they employ. While growers are responsible for providing a safe work environment, their perceptions of the health risk of pesticides influence the amount and quality of safety training and protection they offer workers. One study analyzed growers' and cooperative extension agents' perceptions of farmworker pesticide exposure from in-depth interviews conducted with growers and extension agents who work in western North Carolina. Both groups indicated that the danger of pesticide exposure is exaggerated by the media and the public. They felt that workers were at little risk of exposure because they had received training and protective equipment as required by law and because they were not in direct contact with chemicals. Their perceptions were at odds with results of other research indicating that many farmworkers do not receive the required training and do not always utilize protective gear. Linguistic and cultural barriers contribute to these discrepancies in perceptions and must be addressed if measures to reduce farmworker pesticide exposure are to be effective.[49]

Exposure Risks of Families of Farmer Pesticide Applicators

The Agricultural Health Study, a large research project, enrolled approximately 27,000 licensed private pesticide applicators in Iowa and North Carolina. The study determined that families of farmers who are pesticide applicators have unusual opportunities for direct or indirect exposure to pesticides. These exposures have not been well characterized. Many indirect exposure opportunities exist; for example, 21 percent of homes are within fifty yards of pesticide mixing areas, 27 percent of applicators store pesticides in their homes, and 94 percent of clothing worn for pesticide work is washed in the same machine as other laundry. Direct exposure opportunities also occur; for example, 51 percent of wives of applicators worked in the fields in the last growing season, 40 percent of wives have mixed or applied pesticides, and half of children ages eleven and older do farm chores. The heretofore undisclosed extent of the chances for exposure of family members of farmers who are pesticide applicators makes studies of their health important.[50]

The Agricultural Health Study is unique among studies of occupational groups in that the wives of farmers who apply pesticides and their children are included. Since farmers generally live where they work, their families have many chances to come into contact with pesticides and other farm hazards. These family members' exposures are often less like those in the general non-farm population.

Farm Family Exposure

Residents of Iowa were enrolled in a study investigating differences in pesticide contamination and exposure factors between twenty-five farm homes and twenty-five non-farm homes. The pesticides investigated were atrazine, metolachlor, acetochlor, alachlor, 2,4-D, glyphosate, and chlorpyrifos; all were applied to either corn or soybean crops. A questionnaire was given to all participants to determine residential pesticide use in and around their homes. In addition, a questionnaire was administered to farmers to determine the agricultural pesticides they used on their farms and their application practices. Non-agricultural pesticides were used more in and around farm homes than non-farm homes. Atrazine was the pesticide used most by farmers. Most farmers applied pesticides themselves but only ten (59 percent) used tractors with enclosed cabs, and they typically wore little personal protective equipment. On almost every farm, more than one agricultural pesticide was applied. The majority of farmers changed from their work clothes and shoes in the home, and when they changed outside or in the garage, they usually brought their clothes and shoes inside.

Applying pesticides in tractors with open cabs, not wearing PPE, and changing from work clothes in the home may increase pesticide exposure and contamination. Almost half of the sixty-six farm children under sixteen years of age were engaged in some form of farm chores, with six (9 percent) potentially directly exposed to pesticides. Only two (4 percent) of the fifty-two non-farm children under sixteen had chores, and none were directly exposed to pesticides. Farm homes may be contaminated with pesticides in several ways, resulting in potentially more contamination than non-farm homes, and farm children may be directly exposed to pesticides through farm chores involving pesticides. In addition to providing a description of pesticide use, the data will be useful in evaluating potential contributing factors to household pesticide contamination and family exposure.[51]

Latino Farmworker Pesticide Exposure Perceptions and Beliefs

While a majority of farmworkers in the United States are Latino, few educational materials used in pesticide safety training take their pesticide exposure perceptions and beliefs into consideration. Research delineates several major themes that reflect farmworkers' assumptions about pesticide exposure. One theme reflects the widespread perception that susceptibility to the effects of pesticides is highly individualized; some people are sensitive and experience ill effects, and others are more resistant. Another theme indicates that farmworkers are concerned with the immediate or acute effects of exposure. Very few are aware of potential long-term consequences of pesticide exposure, and none link these to chronic exposures or to residues. Tied to this theme is the belief that the skin acts as a barrier to exposure, rather than as a permeable membrane through which pesticides can be absorbed. Latino farmworkers have their own indigenous health belief system; based on this system, workers often delay washing and showering so as not to apply water, which is cool, to their body when it is hot from rigorous activity. Farmworkers are divided on whether pesticides are potentially dangerous. Some believe that the chemicals are not dangerous to humans, and that they hurt only insects or weeds. While some state that farmers would not use chemicals if they were dangerous to farmworkers, others contend that farmers have no regard for their health. Finally, farmworkers often think they have no control over workplace pesticide safety, and feel unable to adhere to safety rules. These beliefs must be addressed in pesticide safety education if the burden of pesticide exposure among farmworkers is to be reduced significantly.[52]

Adolescent Latino Farmworkers' Pesticide Knowledge and Risk Perception

While exact figures are not known, it is estimated that adolescents make up 7 percent of Latino farmworkers in U.S. agriculture. These young workers may be at increased risk for the toxic effects of environmental exposures encountered during their work. Furthermore, language barriers and health perceptions, similar to those of adult farmworkers, may influence the risk perceptions of this population.

A cross-sectional survey of migrant adolescent farmworkrers was conducted in 1998 to investigate their work practices, health beliefs, and pesticide knowledge. The large majority of the adolescents in the sample were from Mexico, and 36.3 percent spoke primarily indigenous languages. Many of the adolescents (64.7 percent) were traveling and working independent of their parents. Few of the adolescents reported having received pesticide training; however, 21.6 percent of the group reported that their current work involved mixing and/or applying agricultural chemicals.

The results of this study indicate a need for improved pesticide training for youth farmworkers, with specialized education efforts directed toward minorities who speak indigenous languages. Special attention is merited toward adolescent farmworkers who report that their work includes mixing or applying pesticides. As the number of adolescent farmworkers increases in the United States and the characteristics of the migrant influx continue to change, culturally and developmentally appropriate instruments are needed to adequately assess the health beliefs and protective practices of this population.[53]

Another study reinforced the findings of the above investigation. Of 460 hired farmworkers in Washington state who were interviewed, 89 percent did not know the name of a single pesticide to which they had been exposed, and 76 percent had never received any information on appropriate protection measures.[54]

Comments and Concerns of New York State Growers

The following are comments of growers on their surveys. They fall into two categories: public perception or pesticide regulations.

Public Perception

- "The public needs to be educated that they don't have to be afraid of a cornfield. People assume a cornfield is a hazardous waste area. Landlords and neighbors think we are all poisoning the land and that farmers are indiscriminate polluters."

- "Corn is a great crop to grow—easily mechanized, good feed for cattle, high energy source. On the other hand, it is very expensive to grow. The use of chemicals is an absolute must, but growing concerns with the environment has made nonfarmers and consumer groups wary. We need a higher level of education to these groups to help alleviate their fears."

- "Don't forget we as farmers have families. We are concerned about health issues as we are more exposed to chemicals than the consumer is. Our wives and families buy all the groceries at the same stores other people do. We want and deserve good information on which we base our decisions."

- "We need all the products and tools we can get in order to ultimately use less total pesticide. I get very tired of some environmental groups thinking farmers spray pesticides on crops just because they have nothing else to do, or worse yet, don't know any better. We don't put $4,000 or more in a sprayer and then go apply it unless it is absolutely necessary."

Regulations/Certification/Pesticide Use

- "I don't feel we farmers are being treated fairly in regards to the use of pesticides. We now must take tests in order to purchase and use chemicals and then go to some of the 'foolish' meetings in order to get 'points' to get our recertification. The meetings I have attended give me the impression people feel we are not applying the chemicals properly or do not give a hoot about the environment. If they only knew how much we have invested in these chemicals and machinery. We farmers have enough common sense to know not to spray when it's windy, not near open ditches, streams, and ponds, and certainly not near the neighbors' houses where we rent land."

- "I personally feel that there is too much blame on the farmers who try to do everything they can do to be safe with their own and everyone else's lives. I do not like how the average person can go to a department store and buy a pesticide with no training, apply it by dumping it, instead of spraying. Then they blame other people for the problems of the environment. We, the farmers, do not have the money to throw around and waste with not having the training to apply it correctly."

- "More restrictions should be implemented to household and lawn care products. They pose serious environmental problems."

- "I would like to see New York state have quicker approval of newly released pesticides—ones that the EPA and other states have already approved. Each additional agency that requires approval before use only drives the corn growers' cost up for the pesticides, and for crop production, such that New York state, which is a marginal corn production area, will eventually have fewer and fewer farmers producing. I am very concerned about the re-evaluation of atrazine. If it is not reapproved, there will be no replacement for this broad-spectrum and economical herbicide."

- "New York state corn producers have to be competitive with Midwest corn growers who have a much wider spectrum of chemical controls that are priced more competitively. Chemical companies don't register in New York because of delays in registration. Atrazine products and others are more cost effective than many alternatives."

- "To grow food and crops we need the option of chemicals that are safe for all users and the environment. When we lose good products that do a good job, it will cost everybody lots of money. We must read and apply according to the label to protect the environment, people, and products."

- "In New York state, farmers are dying of taxes and other expenses. Other states bordering New York have more access to less-costly pesticides. Atrazine is one of the few chemicals we can use to control small problems with lower cost per acre."

- "With the demise of atrazine formulations, the growing of corn will be very expensive."

- "We need access to newer, safer chemicals. I would not remove older ones from the market, however. We need to be able to rotate chemicals in order to avoid resistance problems."

- "Our main concern in using pesticides in growing crops are the threats of fines and liabilities. Even though we use practical precautions, we are aware things possibly can go wrong. We are very mindful of the effects to persons and the environment, and do our utmost to be careful. It seems that the control agency could be more understanding and helpful in the products they license for the public to use. After all, they and the government make the decision to put these pesticides on the market. By their very act of licensing, they are the ones who introduce hazardous material into the environment."

Though these comments originate from only one state, New York, in all likelihood they are representative of the viewpoints of growers nationwide. The remarks indicate an overall awareness of the problems surrounding pesticide use. Costs were mentioned several times as were the burden of regulations and the public's misperceptions of farm problems. Also of interest were the remarks supportive of the use of the herbicide atrazine, mainly on the basis of its relatively low cost. Its dangers to health, as has been noted earlier, seem to have been ignored.

A Dissenting Voice

The remarks of a fruit grower in Arizona, unlike those of the New York farmers, seem to overlook the possibility of long-term chronic effects of pesticide use, as he emphasizes the benefits of safety precautions during chemical usage: "I've got three very, very healthy kids and it's sort of ludicrous to think they're all going to die because they were exposed to pesticides," says Bill Spencer, who has spent his life raising lemons, tangelos, and grapefruit in Yuma, Arizona. "Farmers are trained in safe application of pesticides. I think there's probably no more family-oriented people in the world than farmers and they're not about to put their children at risk." Further, he says, "I don't think the NRDC [National Resources Defense Council, a leading environmental organization] is aiming their material at farmers. I don't know any stupid farmers out there."[55]

Spencer's comments, on the whole, seem sincere. But one wonders if they would withstand scientific scrutiny in light of information from the EPA's Office of Toxic Substances, to the effect that scientists estimate that everyone alive today carries within his or her body at least 700 chemical contaminants.[56]

Integrated Pest Management: A Mindset

The acronym IPM (Integrated Pest Management) originated in 1967, just a few years after Rachel Carson published *Silent Spring*, the book that unveiled the dangers of pesticides. Today, IPM is often considered cutting edge. It is used by farmers, government institutions, and others who have learned over the years that chemical controls have their price, including waning effectiveness due to pest resistance, high costs, and immediate and long-term health effects to humans and ecosystems.

IPM is more a mindset or long-term strategy than a specific physical solution to pest problems. It requires a number of steps to be taken to reach pest control goals and to subsequently maintain their outcomes. IPM relies on common sense. The "spray and pray" mentality seeks to eradicate all pests—an impossible goal. IPM seeks to eliminate the root causes of pest problems in order to reduce pest numbers to a tolerable minimum.

How Are Farmers Working to Reduce Pesticide Risks?

Today, many farmers are using integrated pest management techniques to minimize pesticide use. IPM works in harmony with nature by using "good bugs," such as ladybugs, to destroy "bad bugs" and other natural control methods. Under IPM, pesticides are used only in limited amounts when pests reach damaging levels, rather than on a routine basis. Many pesticides now being developed use biological or natural substances in the environment to help destroy pests. Research in plant breeding continues to develop heartier, more pest-resistant crops.

The objective of IPM is not to eliminate a pest but to reduce its population to levels that no longer pose an economic threat to plants and animals. IPM is not an "organic" or non-pesticide approach to pest control. Organic producers also use IPM and certain approved pesticides to protect their crops and livestock. One of the main goals of IPM is to promote the use of effective, less-toxic pesticides only if and when necessary. It is a decision-making process that supports a balanced approach to managing crop and livestock production systems. The goal is effective, economical, and environmentally sound suppression of pests, including insects and mites, plant diseases, weeds, and problem wildlife.

The concept of IPM evolved in response to problems caused by an overreliance on chemical pesticides. Some of these problems are development of well-known pest resistance, elimination of natural enemies of pests, outbreaks of formerly suppressed pests, hazards to non-target species, and environmental contamination.

Most growers, if asked (and especially if asked by a concerned consumer), would state that they practice IPM. Most would say that they use pesticides only when necessary and that they are good stewards of the environment. If this is true, perhaps the public should understand better and hear more about the practice of IPM.[57]

Loopholes and Amendments

Pesticide regulations are full of loopholes. Many pesticides in use today were registered using old test protocols and have not yet been reevaluated under current standards. Pesticide manufacturers perform or fund pesticide testing, setting up a built-in conflict of interest. Many tests are only conditionally required and are often waived. Tests ignore the multiple pesticides to which people are regularly exposed because they examine only one pesticide at a time.

There are laws to protect farmworkers, but for almost every law, there is a loophole. For example, OSHA safety standards apply only to farms that hire at least ten workers. That covers about 471,600 farmworkers nationwide, but excludes an estimated one million who labor on small farms.[58]

The EPA Weakens Protection Standards

Even when tougher laws are passed, they are often watered down later. The EPA enacted WPS in 1974. It required growers (regardless of the number of workers) to provide training and information about pesticides used on crops, protective clothing, waiting periods for reentry into treated fields, and hand-washing facilities in the field. But the EPA amended WPS in 1996. Under the amended standards, workers who had never received pesticide training could work five days in the fields without any information about the dangers. The new standards also reduced the number of days that growers must provide water for hand washing (one gallon for every worker) from one gallon a day to one gallon every seven days for certain pesticides. Two years after the EPA relaxed the standards, skin rashes reported by field workers began to climb. In 1998, the rate was about eleven cases per 10,000 workers. By 2001, the rate had jumped to nearly twenty-seven cases per 10,000 workers, among the highest for any occupation, according to the Bureau of Labor Statistics.[59]

Is Organic the Answer?

Organic agriculture is the oldest form of agriculture on Earth. Farming without the use of petroleum-based chemicals (such as fertilizers and pesticides) was the sole option until after World War II. The war brought with it technologies that were useful for agricultural production. For example, ammonium nitrate used for munitions during the war evolved into ammonium nitrate fertilizer; organophosphate nerve gas production led to the development of powerful insecticides. The technical advances since World War II have resulted in significant economic benefits as well as environmental and social detriments. Organic agriculture seeks to utilize those advances that consistently yield benefits, including new varieties of crops, precision agricultural technologies, and more efficient machinery, while discarding those methods that have led to negative impacts on society and the environment, such as pesticide pollution and insect resistance. Instead of using synthetic fertilizers and pesticides, organic farmers utilize crop rotations, cover crops, and naturally based products to maintain or enhance soil fertility. These farmers rely on biological, cultural, and physical methods to limit pest expansion and increase populations of beneficial insects on their farms.

Farmers have been developing organic farming systems in the United States for decades. State and private institutions have also emerged to set organic farming standards and provide third-party verification of label claims. Legislation requiring national organic farming standards was passed in the 1990s. More U.S. producers are considering organic farming systems in order to lower input costs, conserve nonrenewable resources, capture high-value markets, and boost farm income.

Traditionally, organic farms have been smaller than conventional operations. This has been due in part to labor requirements. Organic systems are generally more labor intensive. Studies have found that about 11 percent more labor was required per unit of production in field crops.[60] This difference can be much greater in fruit and vegetable crops, and farm size may be limited accordingly. However, technological innovations in organic horticultural production are helping to narrow the gap. Organic systems are also more information intensive, requiring additional management time in planning, pest scouting, and related activities. For this reason, organic management can be preferable if a farm is not too large. However, the notion that organic systems are only possible on very small farms is a false one. Studies conducted by both Washington University in Washington state and the Department of Agriculture confirmed this. Given the range of acceptable technologies available, organic agriculture can be sized to fit a wide range of farms and enterprises.[61]

A better way to understand organic farming is to hear from the farmers themselves. Here are some quotations from farmers in the Midwest:

- "Why is it that when somebody gets deathly sick with cancer or something, a doctor recommends that they go on an organic diet? I think that all these people know that there is a difference."

- "If you get on a chemical system, the only way you can keep going is to keep adding more and more powerful chemicals; if you get on an organic system, it will perpetuate itself. You don't need to keep adding more and more fertilizer because it is a natural system."

- "If you want to be certified organic you have to demonstrate and have a plan on how you're going to farm and how you're going to produce these crops without using all these chemicals."

- "The nice part about organic is that it's economically viable, and the reason is that you don't have to spend a lot of money, because the good lord designed the cycles of nature in order to do it itself."

- "You can get instant results from chemicals and there is no doubt they will work. They are short-term solutions, but they don't solve any long-term problems."

- "It just got to the point where we didn't think we wanted to use and handle the chemicals. We decided that we could farm without them and we thought we'd give it a try. We just thought it would be a better way of doing things, and we thought we could reduce the cost of production plus increase the quality of what we produce."[62]

Evolution of Organic Agriculture

Organic farming embodies the elements of a sound agriculture—traditional practices that have been proven over time. In fact, a convenient working definition for organic agriculture is "good farming practice without using synthetic chemicals." This working definition distinguishes organic farming from the general milieu of agriculture that existed in the pre-chemical era, much of which was exploitative and unsustainable. Organic farming was never intended to be "throwback" or a regressive form of agriculture.

Rachel Carson's book *Silent Spring* was one of the key documents that gave birth to environmental consciousness in the 1960s and 1970s. Carson was the first to introduce the general public to the concept of persistent bioaccumulative toxins—lasting substances that move readily from land to air and water, and that can build up in the food chain to levels that are harmful to both humans and the environment. These compounds can be found everywhere—in herbicides, pesticides, insecticides, detergents, and cosmetics.

Environmentalists and others found an alternative to pesticides and industrial agriculture, in organic farming. Not only was it an approach that did *not* use synthetic pesticides, it also had an attractive countercultural name that grew to signify a philosophy of living as well as a method of farming. While *Silent Spring* and the environmental movement were not about organic farming per se, they brought it to public consciousness on a vast scale. It is not uncommon, in fact, for some writers to suggest that organic agriculture began with Rachel Carson's book. Though this assertion is untrue, the book clearly played a major role in stimulating industry growth and in altering public perceptions. From the mid-1960s onward, organics was increasingly identified with pesticide issues. It became the idealized alternative for providing clean, healthy food and environmental protection.[63]

As the organic industry continues to grow and evolve, it faces many challenges, including the consequences of its own success. Economic opportunities invite new players into the marketplace; they may have little interest in sustainability or the positive social benefits associated with organics. Noted rural sociologist William Heffernan has touched on this matter. He has gained considerable attention in recent years for his insightful analyses of the causes and social consequences of the increased concentration and corporate control of the U.S. food system. He expressed the following regarding organic farming:

"We are beginning to realize that up to this point we believed that organic was synonymous with family farms and we are finding out that is changing. In fact, the organic is going to continue to grow. That doesn't mean that it is going to support family farms the way it has in the past. With the whole organic movement, we assumed that the social would go along with environmental movement, and what we are finding out is no, that is not necessarily true, and even what we do environmentally is questionable."[64]

As the volume and variety of organic products increases, the viability of the small-scale organic farm is at risk, and the meaning of organic farming as an agricultural method is ever more easily confused with the related but separate areas of organic food and organic certification. For that matter, it remains to be seen whether certified organic farming will survive its own success and continue as a socially and environmentally responsible alternative, or merely become a parallel production system based on minimal compliance to standards.

The Safety Factor

Does government registration mean pesticides are safe? By law, pesticides are regulated by the EPA so that they will not generally cause unreasonable adverse effects on the environment. Pesticide residues on food must be safe with a reasonable certainty that no harm will result from aggregate pesticide exposure. But does this mean that pesticides, by a common-sense definition, are safe? No.

Probably the simplest way to evaluate whether registration means pesticides are safe is to examine recently registered pesticides to see if they meet an acceptable level of safety. As newly registered pesticides, they should adhere to all current standards.

The EPA evaluated nineteen conventional pesticides registered since 1997 and found that most of them pose substantial hazards. Seven cause cancer and six lead to genetic damage. One induces miscarriages, one results in birth defects, one brings on cataracts, one produces bone marrow abnormalities, two are neurotoxic, and one causes both liver and kidney damage. Eight are toxic to fish, five to juvenile fish and three to adult fish. Five have characteristics of groundwater contaminants. Two are highly toxic to oysters, and one to shrimp.[65]

Clearly, these pesticides are far from safe by any common-sense definition. In addition to the known negative effects of pesticides on human health and the environment, there are many effects that are poorly understood. In the United States, decisions on the regulation of pesticides are not based on whether they are safe, but are made via cost-benefit analyses, whereby the financial benefits to industry outweigh the cost to society in human health and environmental damage. Consumers and taxpayers are the ones who pay the price.

The Regulatory Authorities Myth

The greatest myth is that government regulatory authorities ensure agricultural poisons are used safely and cause no adverse health or environmental problems. History shows that regulatory authorities have consistently failed to prevent the contamination of the environment and threats to human health by products previously said to be safe. Think of asbestos, lead, mercury, dioxins, PCBs, DDT, dieldrin, and other persistent organic pollutants (POPs). These products were not (and are still not in many cases) withdrawn until decades after solid scientific evidence demonstrated their damage. Regulatory authorities in the United States (and for that matter, the entire world) seem to be ignoring a large body of published science showing that the current methods of determining the safety of the agricultural poisons are grossly inadequate.

The Pesticide Myth

For fifty years, farmers and the general public have been told that chemical pesticides are essential for modern farming and to feed the world's population. This simply isn't true. Pesticides weaken the ecosystem that has sustained human agriculture for thousands of years, damaging soil microbes and eliminating beneficial insects and predators. In addition, pests continually mutate to become pesticide resistant. Despite a tenfold increase in insecticide use in recent years, research has demonstrated a proliferation of pests.

Gardeners, farmers, and foresters need to return to tried-and-true pest control methods such as crop rotation, companion planting, and biological controls. Integrated Pest Management, which uses fewer toxic chemicals, and then only infrequently, is the pest control of the future. Ecological methods of pest control must replace our overdependence on chemicals which now threatens us all. Numerous studies show that IPM can save significant amounts of money for farmers while helping them protect their health and the environment.

The most important and urgent step needed to reduce exposure is eliminating use of those pesticides which endanger the health and well-being of farmworkers. Their experiences reveal that even pesticide applications that follow the letter of the law can result in exposure or illness. Phasing out the use of the most dangerous pesticides—those that cause cancer or reproductive harm, or are extremely toxic to the nervous system—would represent a tremendous step toward a more sustainable, healthier, and more humane agricultural system.

Notes

1. Remarks of Laura Caballero, Salinas, California, public meeting, July 1996.

2. Caroline Cox, "Working with Poisons on the Farm," *Journal of Pesticide Reform* 14 (3) (Fall 1994): 2–5.

3. S. A. Briggs, *Basic Guide to Pesticides: Their Characteristics and Hazards* (Washington, D.C.: Rachel Carson Council, 1992).

4. *Overview of America's Farmworkers* (Buda, TX: National Center for Farmworker Health, Inc., n.d.): 3.

5. R. Mines, S. M. Gabbard, and G. Stewart, *A Profile of U.S. Farmworkers: Demographics, Household Composition, Income and Use of Services,* based on data from the National Agricultural Workers Survey (Washington, D.C.: U.S. Department of Labor, 1997).

6. K. Mobed, E. B. Gold, and M. B. Schenker, "Occupational Health Problems Among Migrant and Seasonal Farmworkers," *Western Journal of Medicine* 157 (1992): 367–373.

7. Phil Kellerman, *Some Facts and Figures* (Stephentown, NY: The Harvest of Hope Foundation, October 20, 2003): 1.

8. Testimony of Arturo S. Rodriguez, President, United Farm Workers of America, AFL-CIO, before the Subcommittee on Employment, Safety, and Training, Committee on Health Education, Labor, and Pensions, U.S. Senate, February 27, 2002.

9. General Accounting Office, *Child Labor in Agriculture: Characteristics and Legality of Work* (Washington, D.C.: U.S. General Accounting Office, 1998): 20.

10. U.S. Environmental Protection Agency, *Basic Principles of the Worker Protection Standard* (July 6, 1999).

11. W. S. Pease, R. S. Morello-Frosch, D. S. Albright, A. D. Keyle, and J. C. Robinson, *Preventing Pesticide-Related Illness in California Agriculture: Strategies and Priorities* (Berkeley, CA: California Policy Seminar, 1993).

12. Rebecca Clarren, "Harvesting Poison," *High Country News* (September 29, 2003).

13. U.S. General Accounting Office, *Pesticides: Improvements Needed to Ensure the Safety of Farmworkers and Their Children* (Washington, D.C.: U.S. General Accounting Office, March 14, 2000): 19.

14. "Occupational Exposure Assessment," *Bayer Crop Science* (West Lafayette, IN: Purdue University Cooperative Extension Service, 2003).

15. Don Villarejo and Celio Prado, "WPS Unknown To Farmworkers," *On-Line News* (Davis, CA: University of California Health and Safety Center, Winter 1997): 4.

16. Mark Dwortzan, "Caution: Work Can Be Hazardous to Your Health," *Harvard Public Health Review* (Fall 2003): 2.

17. National Resources Defense Council, "Trouble on the Farm: Growing Up with Pesticides in Agricultural Communities" (October 1998): 4.

18. "Farm Worker Illness Following Exposure to Carbofuran and Other Pesticides—Fresno County, California 1998," *Morbidity and Mortality Weekly Report* 48 (6) (February 19, 1999): 113–116.

19. J. L. Runyan, *Summary of Federal Laws and Regulations Affecting Agricultural Employers* (Washington, D.C.: U.S. Department of Agriculture, 2000).

20. Penelope A. Fenner-Crisp, "Pesticides—The NAS Report: How Can the Recommendations Be Implemented?" *Environmental Health Perspectives* 103 (Suppl. 6) (1995): 159–162.

21. Andrea Helm, "EPA Waste Policy Threatens Health," *GREEN LINE* 4 (9) (June 1991): 116–118.

22. Subtitle C of Public Law 94-580, Resource Conservation and Recovery Act of 1976.

23. Stephen Lester, "Secret Ingredients in Pesticides: Toxic Waste," *EVERYONE'S BACKYARD* 9 (5) (October 1991): 7–8.

24. Burt McMeen, "Groups Petition for Protection of Children Living on Farms, Citing Exposure Studies," *Toxics Law Reporter* 13 (24) (November 11, 1998): 752.

25. M. Schenk, S. M. Popp, and A. W. Neale et al., "Environmental Medicine Content in Medical School Curricula," *Academic Medicine* 71 (5) (1996): 27–29.

26. J. N. Thompson, C. A. Brodkin, K. Kyles, W. Neighbor, and B. Evanoff, "Use of a Questionnaire to Improve Occupational and Environmental History Taking in Primary Care Physicians," *Journal of Occupational and Environmental Medicine* 42 (2000): 1,188–1,194.

27. William M. Simpson Jr. and Stanley H. Schuman, "Recognition and Management of Acute Pesticide Poisoning," *American Family Physician* 65 (2002): 1,599–1,604.

28. May 1999 Meetings of the Education and Practice Workgroups (Washington, D.C.: Environmental Protection Agency, Office of Pesticide Programs).

29. National Occupational Research Agenda, *Factors in Determining the Occupational Health Status of Hired Farmworkers* (Washington, D.C.: National Institute of Occupational Safety and Health, n.d.).

30. Ibid.

31. U.S. General Accounting Office, *Pesticides: Improvements Needed.* Op. cit. 11–12.

32. Center for the Economy and the Environment, *Evaluating Environmental Progress: How EPA and the States Can Improve the Quality of Enforcement and Compliance Information* (Washington, D.C.: The National Academy of Public Administration, June 2001).

33. Center for Disease Control and Prevention, "Organophosphate Insecticide Poisoning Among Siblings—Mississippi," *Morbidity and Mortality Weekly Report* (33) (1984): 592.

34. R. J. Zweiner and C. M. Ginsburg, "Organophosphate and Carbamate Poisoning in Infants and Children," *Pediatrics* 81 (1998): 121–126.

35. "GAO Says EPA Needs to Act on Farm Pesticides, Children," *Toxic Chemicals Litigation Reporter* 18 (1) (June 2, 2000): 8.

36. "Officials Want to Improve Farmer Medical Care," *BC Cycle* (May 30, 1990).

37. Susan Gabbard, Rick Mines, and Andrea Steirman, *A Profile of U.S. Farm Workers: Demographics, Household Composition, Income and Use of Services* (Washington, D.C.: U.S. Department of Labor, Office of Policy, 1997): 1.

38. J. E. Davies, V. H. Freed, and H. F. Enos et al., "Reduction of Pesticide Exposure with Protective Clothing for Applicators and Mixers," *Journal of Occupational Medicine* 24 (1982): 464–468.

39. M. J. Coye and J. A. Lowe et al., "Biological Monitoring of Agricultural Workers Exposed to Pesticides: II: Monitoring of Intact Pesticides and Their Metabolites," *Journal of Occupational Medicine* 28 (8) (1986): 628–636.

40. "Farmworker Advocates Seek Phaseout of Neurotoxic Pesticides Based on Biomonitoring Data," *Insider eJournal* 2 (4) (March 1, 2005).

41. "Part II—Chemical Management," *Pennsylvania Tree Fruit Production Guide* (University Park, PA: Penn State College of Agricultural Sciences, 2006–2007).

42. Jocelyn Kaiser, "Endocrine Disrupters: Synergy Paper Questioned at Toxicology Meeting," *Science* 275 (5308) (March 28, 1997): 1,879–1,880.

43. Department of Agriculture, *Agricultural Chemical Usage, 2001 Field Crops Summary* (Washington, D.C.: U.S. Department of Agriculture, 2002).

44. Rex Dufour, *Biotensive Integrated Pest Management (IPM)* (Fayetteville, AR: National Sustainable Agriculture Information Service, July 2001): 1.

45. Patricia S. Muir, *B1301 Human Impacts on Ecosystems* (Corvallis, OR: Oregon State University, Fall 2005).

46. *Trouble on the Farm: Growing Up with Pesticides in Agricultural Communities* (Washington, D.C.: National Resources Defense Council, October 1998): 53.

47. S. H. Zahm, M. H. Ward, and A. Blair, "Pesticides and Cancer," *Occupational Medicine State of the Art Review* 12 (2) (1997): 269–289.

48. R. Zwiener and C. Ginsburg, "Organophosphate and Carbamate Poisoning in Infants and Children," *Pediatrics* 81 (1988): 1,211–1,226.

49. Pamela Rao, Thomas A. Arcury, Sara A. Quandt, and Alicia Doran, "North Carolina Growers and Extension Agents' Perceptions of Latino Farmworker Pesticide Exposure," *Human Organization* 63 (2) (2004): 151–161.

50. B. C. Gladen, D. P. Sandler, and S. H. Zahm et al., "Exposure Opportunities of Families of Farmer Pesticide Applicators," *American Journal of Industrial Medicine* 34 (6) (December 1998): 531–537.

51. B. Curwin, W. Sanderson, S. Reynolds, M. Hein, and M. Alavanja, "Pesticide Use and Practices in an Iowa Farm Family," *Journal of Agricultural Safety and Health* 8 (4) (2002): 423–433.

52. Thomas A. Arcury, S. A. Quandt, P. Rao, and G. B. Russell, "Farmworker Pesticide Exposure Perceptions and Beliefs: Using Cultural Knowledge to Improve Safety Education," paper presented at Society for Occupational and Environmental Health 2002 Conference, July 9, 2002, Bethesda, MD.

53. L. A. McCauley, D. Sticker, and C. Byrd et al., "Pesticide Knowledge and Risk Perception Among Adolescent Latino Farmworkers," *Journal of Agricultural Safety and Health* 8 (4) (2002): 397–409.

54. Michelle Mentzer and Barbara Villalba, *Pesticide Exposure and Health: A Study of Washington Farmworkers* (Seattle, WA: Evergreen Legal Services, March 1988): 28–29.

55. Michael Fumento, "City Slickers Off Target in Pesticide Report," *Idaho Statesman*, December 15, 1999.

56. J. Onstot, R. Ayling, and J. Stanley, *Characterization of HRGC/MS Unidentified Peaks from the Analysis of Human Adipose Tissue, Volume 1: Technical Approach* (Washington, D.C.: U.S. Environmental Protection Agency, Office of Toxic Substances, 1987).

57. *Agricultural Pesticides: Management Improvements Needed to Further Promote Integrated Pest Management*, GAO Report to the Chairman, Subcommittee on Research, Nutrition, and General Legislation, Committee on Agriculture, Nutrition, and Forestry, U.S. Senate, August 17, 2001.

58. Christine Stapleton, "Pickers Wade in Pesticides, but Training and Oversight Are Lax," *Palm Beach Post*, December 7, 2003.

59. Ibid.

60. R, Klepper, W. Lockeretz, and B. Commoner et al., "Economic Performance and Energy Intensiveness on Organic and Conventional Farms in the Corn Belt: A Preliminary Comparison," *American Journal of Agricultural Economics* 59 (1) (February 1977): 1–12.

61. William Lockeretz, Georgia Shearer, and Daniel Kohl, "Organic Farming in the Corn Belt," *Science* 211 (6) (February 1981): 540–547.

62. H. S. James Jr., "Conversations with Missouri Corn and Soybean Producers," *Journal of Agricultural Safety and Health* 11 (2): 239–248.

63. Rachel Carson, *Silent Spring* (Boston, MA: Houghton Mifflin Co., 1962).

64. Donella Meadows, "Our Food, Our Future," *Organic Gardening* (September–October 2000): 53–59.

65. "Does Government Registration Mean Pesticides Are Safe?" *Journal of Pesticide Reform* 19 (2) (Summer 1999).

Three

Pesticides in Food

The destiny of nations depends on the manner in which they feed themselves.

—Jean Anthelme Brillat-Savarin, *The Physiology of Taste* (1825)

Food Safety

The safety of food is an age-old concern. Early civilizations adopted laws that punished sellers of tainted food. In this country, before food safety became a responsibility of the federal government, every state had laws prohibiting the sale of food that contained poisonous substances. The modern scientific and legal instruments available to the U.S. Food and Drug Administration (FDA) and allied agencies have improved regulation and advances in food preparation, preservation, and storage, contributing to a safer food supply. Even so, there is a belief that contemporary threats to food safety have grown more serious; they surely excite intense public concern.

Supporters argue that pesticide use is necessary to keep the cost of food production low and to maintain an abundant, affordable supply of fruits and vegetables in the market. However, opponents argue that since pesticide-free agriculture has never been tried on a large enough scale, we really do not know if the cost of food production would increase, or by how much. Researchers have studied the profitability of farms that do not use synthetic pesticides and found that results can vary depending on the kind of crop and region of the country.

Meanwhile, the soil is being saturated with poison sprays, which means that many fruits and vegetables absorb the pesticides systemically through their roots. While one can wash some of the poison off the outside of these fruits and vegetables, it has become part of the produce and cannot be completely removed.

Regulation of Pesticides in Food

In the United States there are three government agencies that share responsibility for the regulation of pesticides: the EPA, the Food Safety Inspection Service of the

U.S. Department of Agriculture (FSIS-USDA), and the FDA. It is the responsibility of the EPA to register (that is, approve) and set tolerances if the use of a particular pesticide may result in residues on food. A tolerance is defined as the maximum quantity of a pesticide residue permitted on a raw agricultural commodity. Tolerances impact food safety by limiting the concentration of a pesticide residue allowed on a commodity, and by limiting the type of commodity on which it is allowed. Tolerances are the only tool the EPA has under the law to control the quantity of pesticides on the food we eat.

The FSIS branch of the USDA is responsible for monitoring and enforcing tolerances of pesticide residues on meat, poultry, and certain egg products. The FDA is charged with enforcing tolerances in imported and domestic foods (predominantly fresh fruits and vegetables) and is responsible for enforcing these tolerances. To be able to enforce the EPA-mandated tolerances, both the FDA and state agencies must know the quantity and the type of pesticide residues present in foodstuffs offered for sale.

The FDA's approach to pesticide residue monitoring involves collecting samples of individual lots of domestically produced and imported foods as close as possible to their point of entry into the distribution system. Imported samples are collected at their point of entry into the United States. The samples are analyzed for pesticide residues to enforce the tolerances set by the EPA. The FDA approach places emphasis on raw agricultural commodities. These are analyzed in a raw, unpeeled, and unprocessed state. The FDA also analyzes processed foods for pesticide residues. When illegal pesticide residues are found, the FDA can impose various sanctions, including seizure of the commodity or injunction. For those samples imported into the United States, shipments are stopped at the port of entry if they are found to obtain illegal residues. If there is reason to believe that future lots from a particular foreign grower or geographic region may be in violation during a given season, the FDA can invoke detention without physical examination (called automatic detention). In this case the food will be detained at the port of entry until analysis is complete. The United States imports approximately 15 percent of total domestic consumption of agricultural products, according to the EPA, and pesticides are used in producing and storing many of these imports. Importation of food with a pesticide residue that exceeds its tolerance is prohibited. According to the General Accounting Office (GAO), in 1994 the FDA tested about 1 percent of all imported shipments for pesticide residue levels.

Critics such as the National Association of State Departments of Agriculture (NASDA) have argued that this monitoring rate has been inadequate, making it unlikely that illegal pesticide residues would be detected on imported foods. The FDA has argued in the past that the low sampling rate had understated the effectiveness of its detection program because the agency had concentrated its efforts on those foods and countries likely to be the source of residues and on shippers with a history of violations. This strategy was intended to identify violations more successfully than more frequent but random sampling. However, FDA officials have admitted that inadequate resources were the primary reason that the agency had not tested a larger percentage of imported foods.[1]

The Pesticide Data Program

In the early 1990s surprisingly little was known about the frequency or levels of pesticides in food as actually eaten. Then-existing government data on residues had been collected as part of tolerance enforcement programs and represented residues at the farm gate, prior to washing, shipping, storage, marketing, and preparation. Relatively insensitive analytical methods were used. To improve the accuracy of pesticide dietary risk assessments, in 1991 Congress funded a new USDA program, the Pesticide Data Program (PDP). By design, the PDP focuses on the food consumed most heavily by children. Food is tested, to the extent possible, "as eaten." Banana and orange samples are tested without their peels; processed foods are tested as they come out of cans, jars, or freezer bags.[2]

Pesticide Cause

A number of surveys indicate that pesticide treatments of food after it has been harvested are more likely to leave residues than the treatment of crops during cultivation. Strawberries that need to be shipped over long distances may be treated after harvest with fungicides to prevent fungi or molds. This could be one reason for the higher-than-average frequency of violative residues on strawberries.

Detectable Pesticide Residues

Pesticides used to enhance food production are commonly separated into four different categories: herbicides, insecticides, parasitic worm killers, and fumigants. Some pesticide-use estimates also account for chemicals such as sulfur and petroleum that are registered as pesticides but produced mostly for other purposes. It is estimated that, of the three major categories of pesticides, insecticides are generally the most toxic, followed by herbicides and then fungicides. Parasitic worm killers are the least toxic. These estimates are based on both chronic and acute toxicity scores.[3]

The EPA is charged with establishing maximum allowable residue tolerances for pesticides and the FDA monitors and regulates the U.S. food supply for compliance with these tolerances. In addition, the PDP collects data on pesticide residues on food. For example, foods sampled by the PDP from 1994 to 1998 showed the following trends: Single residue detections on sampled food commodities exhibited a stable trend, ranging from 25 to 27 percent on sampled foods each year. Multiple residue detections on sampled food commodities declined after an initial rise from 36 percent of sampled foods in 1994 to 29 percent in 1998. Overall, detectable residues on sampled foods decreased from 61 percent in 1994 to 55 percent in 1998.[4]

Data Characteristics and Limitations

PDP samples are collected by ten participating states, which represent all regions of the country and half of the national population. Samples are collected near end markets and large chain-store distribution centers.

The PDP's sampling strategy is statistically reliable and allows for realistic estimates of pesticide residues in the total food supply and of consumer exposure to them. Data are available annually and are reported by the food product and the pesticide for which the food product was tested.

Detectable and Violative Pesticide Residues

In the United States and other developed countries, including Japan and the nations of Western Europe, the majority of pesticide applications represent herbicides, which tend to have lower acute toxicities than insecticides. However, in developing countries, the situation is reversed. In these countries, insecticides are primarily used—often older compounds in the organophosphate and carbamate families known for their acute and chronic toxicities. Because of the potential risks to human health posed by agricultural practices in other countries, it is important to monitor pesticide residues on food imports. The FDA enforces the EPA's pesticide residue tolerances in imported foods. The foods sampled by the FDA from 1993 to 1999 with detectable and violative pesticide residues revealed these trends: Total pesticide residue detections exhibited a reasonably stable trend, ranging from 31 percent to 35.6 percent of sampled food imports. Violative pesticide residue detections also exhibited a reasonably stable trend; however, the 1997–1999 period demonstrated a slight increase, from 1.6 percent to 3.1 percent of sampled food imports.

Data Characteristics and Limitations

The FDA samples both raw agricultural commodities and processed food products. It relies on multiresidue methods (MRMs) that can simultaneously detect a number of different pesticide residues. In 1999 the FDA collected 6,012 food samples representing shipments from ninety-two countries.[5]

Violations or Presumed Tolerance Violations

A violation occurs when a residue is detected that exceeds its tolerance or when a residue is found for which there is no tolerance set for a specific crop. Since 1991, the PDP has tested fruits, vegetables, grains, dairy, and processed products for residues of more than 160 different pesticides. Foods were sampled from 1993 to 1998 that had residues that violated or were presumed to violate (that is, no tolerance level was established for that crop) tolerances. Each year, less than 0.2 percent of all sampled foods had residues that violated established tolerances. Despite a decrease from 1997 to 1998, the percentage of sampled foods presumed to violate tolerances exhibited an increasing overall trend—from 1.31 percent in 1993 to 3.7 percent in 1998.[6]

Pesticide Residues: Reducing Dietary Risks

Recent data on pesticide residues, food consumption, and pesticide use reveal both the sources of consumers' dietary intake of pesticide residues and the benefits of research to develop safe alternatives to pesticide use. Consumers' dietary intake comes from four sources: on-farm pesticide use, post-harvest pesticide use, pesticides used on imported foods, and canceled pesticides that persist in the environment. Post-harvest uses account for the largest share of dietary intake of residues, but canceled and persistent chemicals appear among the highest risk indicators. Thus, research to develop on-farm pest control alternatives will not address all of the sources of these residues. While most pesticide use does not result in detectable residues, higher levels of use do result in higher residues. The geographic source of residues can be identified.[7]

Underreporting Pesticide Residues

Pesticides in children's food are the subject of a report that examined the patterns of food consumption of young children and the monitoring capabilities of the FDA's laboratories. It considered that FDA seriously underreports pesticide residues in the food supply. From 80 percent to 100 percent of residue analyses at five of twelve FDA regional laboratories were not capable of finding 80 percent of pesticides used in agriculture. This investigation, which also considered the exposure of children to pesticides that the EPA considers probable or possible human carcinogens, estimated that, by the age of five, millions of children have already received up to 35 percent of their entire lifetime doses of some cancer-causing pesticides. This pattern is most evident in pesticides used on foods heavily consumed in the first years of life. These include the fungicides captan (35 percent of lifetime risk by age five) and benomyl (29 percent), and the insecticide dicofol (32 percent).

Changing regulatory procedures is not considered the optimum means to move forward. Rather, the report stressed the importance of continuing to eat fresh fruit, vegetables, and other staples. It also advocated reducing the use of pesticides in food production, including: 1) a targeted pesticide risk reduction strategy that will gradually phase out the use of pesticides that present the greatest hazards to children, including all those classified by the EPA as known carcinogens or potential carcinogens; 2) a program of research for agricultural producers to help them develop alternative pest control practices for high-risk pesticide/crop combinations; and 3) steps to expand consumer access and farmers' markets for foods produced with fewer pesticides and that contain no residues, and a voluntary no-detected or ultra-low standard for pesticide residues in food.[8]

Children Differ From Adults

A seminal 1993 report by the National Research Council, *Pesticides in the Diets of Infants and Children*, found that existing EPA standards for setting allowable pesticide residue limits did not consider the unique vulnerabilities and exposures of infants and children.

Most of the data used in regulating pesticides come from food surveys and laboratory tests that don't adequately account for the vulnerabilities of children. It is critical that regulators evaluate differences between children and adults that may affect pesticide-related health risks.

One important difference is in eating patterns, which to a large degree determine a person's level of exposure to pesticides. Infants and young children eat fewer foods and thus consume much more of certain foods per unit of body weight. The fact is that children's eating patterns are inadequately represented in food consumption surveys and measurements of pesticide residues.

In most food consumption surveys, data on food intake are grouped by broad age categories, such as one- to six-year-olds. This method obscures rapid changes in diet that occur as children grow. The surveys usually focus on average intake within these broad age groups. This reliance on averages may cause regulators to overlook geographic, ethnic, socioeconomic, and other factors that can affect exposure. Measurements of pesticide residues, which tend to focus on foods eaten by the average adult, underrepresent foods consumed by infants and children.

Another way children may differ from adults is in sensitivity to toxic substances. Little is known about children's sensitivity to pesticides. But data on other toxic chemicals suggest that children may be more sensitive than adults to some pesticides, while being less or equally sensitive to others.

These differences in sensitivity are usually small, generally less than tenfold. Still, the sensitivities of infants and children should be more fully studied than they are now. Most lab tests conducted by pesticide manufacturers to satisfy EPA requirements "are designed primarily to assess pesticide toxicity in sexually mature animals."[9]

The Pre-FQPA Situation

The long-standing debate concerning pesticide regulation made one thing evident to advocates on all sides of the controversy—the laws regulating pesticide use and pesticide residues in food had to be reformed. Congress responded with the passage of Food Quality Protection Act (FQPA) of 1996, amending the Federal Insecticide, Fungicide, and Rodenticide Act (FIFRA) and the Federal Food, Drug, and Cosmetic Act (FFDCA). This bipartisan effort passed unanimously in both the House and the Senate.

Pre-FQPA pesticide standards were based on healthy adult males, rather than considering the entire population, including children and other vulnerable groups. Children's unique susceptibilities and usually higher exposures to pesticides were not taken into account. For example, pre-FQPA testing procedures did not, for the most part, adequately address the toxicity and metabolism of pesticides in newborn and adolescent animals or the effects of exposure during early developmental stages and their effects later in life. One analysis of several pesticides suggested that "the reference dose (the dose of a non-cancer toxicant at which no health effects are likely) may be exceeded by thousands of children daily."[10]

Before the FQPA, pesticide tolerances were set balancing health against agricultural economics and other considerations. The FQPA makes children's health the legal standard for setting tolerances. The standards set must assure "a reasonable certainty of no harm" to children's health, a new and more protective requirement. Pre-FQPA pesticide regulation did not include answering basic questions such as:

1. Could the pesticide affect a child's behavior, learning, or memory?

2. Does it affect the developing nervous system, especially since many pesticides work by short-circuiting the nervous system, and if so, how?

3. What is the impact of multiple pesticide exposures, as occurs daily in real life?

4. What is the pesticide's effect on the immune system?

5. Does the pesticide disrupt hormone systems?[11]

The Food Quality Protection Act of 1996

The statute removed the Delaney Clause and revised the approach to how risk assessments are conducted, including a specific mechanism to consider risks to children. The law requires an explicit determination that tolerances are safe for children, includes an additional safety factor of up to tenfold to account for uncertainty in data, and requires consideration of children's special sensitivity and exposure to pesticides.

These amendments fundamentally changed the way the EPA regulates pesticides, based on a new standard of "reasonable certainty of no harm" that must be applied to all pesticides used on foods. For more than two decades there were efforts to update and resolve inconsistencies in the two major pesticide statutes, but consensus on necessary reforms remained elusive. The FQPA represented a major breakthrough, amending both major pesticide laws to establish a more consistent, protective regulatory scheme grounded in sound science. It mandates a single, health-based standard for all pesticides in all foods; provides special protections for infants and children; expedites approval of safer pesticides; creates incentives for the development and maintenance of effective crop protection tools for American farmers; and requires periodic reevaluation of pesticide registrations and tolerances to ensure that the scientific data supporting pesticide registrations will remain current in the future.

Revising Tolerance-Setting Criteria for Pesticide Residues in Food

A key issue in the 104th Congress was whether to revise the so-called zero-risk standard of the Delaney Clause (FFDCA, Section 409), which prohibits the addition of potentially cancer-causing substances to foods. The application of the Delaney Clause to pesticide residues was criticized for being unscientific and for creating a confusing and inconsistent set of standards for safety, depending on whether a pesticide was on raw or processed food and whether it was a carcinogen or not. Critics of pesticide regulation under Delaney maintained that it was unscientific because very low pesticide residues pose no significant risk to health. Technology is now sophisticated enough to detect extremely small amounts of pesticides in food, in some case levels in parts per trillion. Thus, food industry representatives claimed that rigid enforcement of Delaney (that is, banning any measurable pesticide concentration) stifled research and development of new pesticides that might have been safer than products already on the market. Critics noted that many foods contain natural carcinogens, which were not regulated under Delaney, which may be more concentrated and more potent than pesticide chemical residues. Critics also said that residues might even have resulted from pesticide use to control fungi or bacteria that produce natural carcinogens. In addition, they claimed that in some cases, the distinction between raw and processed foods made no sense. The absolute amount of pesticide in a food before and after processing might be the same, yet a tolerance could be set for the residue in raw food and prohibited for the residue in processed food, because the residue had concentrated relative to the total food weight due to drying or other processing.

Proponents of the Delaney Clause argued that the public does not want to be exposed to carcinogenic pesticides in food no matter how small the risk. With regard to naturally occurring carcinogens in food, supporters argued that federal agencies would not readily assess and reduce that risk, especially since natural anti-carcinogens often are found in the same food as the carcinogens. To reduce the overall cancer risk, therefore, they believed the federal government should minimize chemical pesticide residues in food.

The Delaney Clause also was problematic, according to some, because it required regulators to treat potentially carcinogenic pesticides more stringently than pesticides that may exert other health effects. This situation set up a paradox: by stringently regulating carcinogens, the "zero-risk" standard Section 409 may have reduced the safety of some foods. Section 409 allowed approval of pesticide residues that posed greater risks than residues of carcinogens which Section 409 did not permit, because many registered pesticide products have health effects other than cancer. For this reason, the National Research Council of the National Academy of Sciences recommended in 1987 that all pesticide residues in food, whether raw or processed, should be regulated on the basis of a consistent "negligible-risk" standard.[12]

FQPA Mandates

A key expressed purpose of the FQPA is to coordinate pesticide registration under FIFRA with FFDCA tolerances to ensure that any pesticide approved for use on food would leave only a "safe" residue. The FFDCA, as amended by the FQPA, defined "safe" to mean that the EPA has determined there is "a reasonable certainty that no harm will result from aggregate exposure ... including all anticipated dietary exposures and all other exposures for which there is reliable information." The FQPA directed the EPA to reevaluate all existing tolerances for food-use pesticides against this safety standard: 33 percent by August 3, 1999, 66 percent by August 3, 2002, and 100 percent by August 3, 2006. The FQPA required the EPA to consider tolerances for the riskiest pesticides first.

If the EPA finds that residues of a pesticide used on food may pose a risk greater than the FQPA allows, the act requires a change in the FFDCA tolerance level, as well as in the FIFRA registration (that is, the product label) to restrict the number or manner of approved pesticide uses, and so to reduce human exposure to a "safe" level. In assessing the risk of pesticide residues allowed by a tolerance, the FQPA requires the EPA to consider: 1) children's exposure to pesticides and susceptibility to health effects; 2) potential disruptive effects on endocrine systems; 3) potential effects of in utero exposure; 4) aggregate risk from all sources and through all routes of exposure; and finally, 5) cumulative risks due to exposure to all pesticides with similar toxic effects, or what is known as a "common mechanism of toxicity."

FQPA Implementation

The EPA has worked to implement the FQPA. Pesticide producers and users want assurances that the EPA will evaluate the risks of popular pesticides based on real data rather than on worst-case assumptions. Public health leaders and environmental groups want prompt action to reduce risks from pesticides.

Progress Toward Milestones

On the date of FQPA enactment, there were 9,728 residue tolerance levels and exemptions in effect for active and inert pesticide ingredients. The EPA divided these into groups based largely on their relative risks to public health, and published a schedule for reevaluation of tolerances in the *Federal Register* on August 4, 1997. The first group of pesticides subject to tolerance reassessment included: 1) organophosphates, carbamates, and organochlorines; 2) those that are probable and some possible human carcinogens; 3) high-hazard inert ingredients; 4) pesticides that exceed their reference dose (RfD); 5) pesticides that the EPA will be considering for reregistration; and 6) pesticides whose tolerances and exemptions are being revoked.

The EPA reevaluated more than 3,000 tolerances before August 3, 1999, the earliest statutory deadline. The agency asserted that it had achieved its first milestone for food-use pesticide regulations. On August 3, 2002, the EPA announced that it had completed reassessment for more than 6,400 tolerances, including nearly two-thirds of tolerances for foods commonly eaten by children, meeting the second statutory deadline. The EPA has revoked more than 1,900 tolerances. However, critics contend that the EPA has not evaluated the riskiest pesticides, since many of the reevaluated tolerances posed no significant risks to human health; many were for residues on crops that did not occur, because the crops were not treated with the pesticide, or the pesticide was no longer in use.

Data Controversy

A particularly contentious implementation issue revolves around FQPA directives to use "available data" and "reliable data," as well as the FQPA mandate to order testing if the EPA determines that data are "reasonably required to support the continuation of a tolerance or exemption that is in effect ... for a pesticide chemical residue on a food" (FFDCA, Section 408(f)(l)). There is disagreement about what is an appropriate course of action for the EPA when there is insufficient "reliable" data to estimate risks. Pesticide producers ideally would like the EPA to delay estimating risks until reliable data can be collected; public health advocates would like the EPA to estimate risks based on "available" data and to reduce the potential for human exposure to unacceptable risks.

Members of the pesticide industry also want the EPA to "call in" data. Although pesticide producers conduct toxicity testing, and they need not wait for the EPA to order data production, an EPA order provides certain legal and financial protections not otherwise available to those who perform toxicity studies. The EPA's failure to order a data call-in was another issue raised by a lawsuit in June 1999, but the court dismissed this claim. However, the EPA published a call-in notice for specific data on developmental neurotoxicity and pesticide residues on August 6, 1999.[13]

When Congress passed the FQPA, many hailed it as an example of a rational, scientific, and risk-based law that would be good for producers and consumers alike. It established a new standard for food safety that recognized the benefits of pesticide use on food crops, but also guaranteed pesticide residues almost certainly would be harmless.

The EPA claims to be meeting statutory deadlines, but various interest groups have challenged that claim, as well as the EPA's implementation process. As food uses of some popular pesticides have been canceled, some policy makers have argued that the FQPA needs to be amended, or that the EPA needs to be restrained from going beyond what the act requires.

Increased congressional oversight and support for legislative remedies might be expected as FQPA implementation proceeds. In 2003 the EPA was expected to complete a cumulative risk assessment of organophosphate (OP) pesticide exposure and to make final decisions about remaining OP pesticide registrations. The FQPA provides little guidance on how the EPA should weigh one pesticide use against another; the EPA arguably has considerable discretionary power to decide which OP uses to permit and which to eliminate. Some of the agency's decisions almost certainly will be challenged in court, quite possibly both by producers and by environmental or public health interests.[14]

Does the FQPA Really Protect the Consumer?

As noted above, in place of the Delaney Clause, the law established a single, health-based, risk assessment standard for all pesticide residues on food, whether fresh or processed. According to the statute, the EPA must issue a finding that a pesticide residue is "safe" in order for it to be allowed. Safe is defined as "a reasonable certainty of no harm" to consumers, and has a legislative history that defines it as a one-in-a-million risk of cancer over a person's lifetime.

Critics of risk assessment point out that this technique is inherently flawed because it operates on the assumption that regulators can determine an "acceptable" level of harm to inflict on unconsenting people, and because it does not provide a framework for exploring alternatives to pesticides. Additionally, although risk assessment techniques are touted as scientifically sound, they are generally based on limited data and can lead to wide-ranging interpretations regarding what levels of exposure are "safe."

The FQPA requires that the EPA consider cumulative exposures (both dietary and nondietary) to a single pesticide; however, it does not require evaluation of exposure to multiple chemicals that have the same adverse health effect, such as nerve damage. This is significant since a single food can have residues of dozens of pesticides, including multiple carcinogens and endocrine disrupters.

The FQPA also contains provisions that would prohibit states from adopting more protective food safety standards, except where the EPA finds that there are special considerations of merit, such as compelling local conditions. The act also gives the EPA power, in some circumstances, to allow use of pesticides that do not meet the new standards when it considers such use necessary to avoid "a significant disruption in the production of an adequate, wholesome, and economical food supply." In these cases, states will be allowed to establish more stringent standards.

Moreover, in recognizing the need for regulations that specifically address children's vulnerabilities to pesticides, the law omits other vulnerable subgroups. Farmworkers, farmers, pesticide applicators, and chemical factory workers all face distinct hazards from pesticide exposure, hazards that the FQPA does not take into account when setting pesticide tolerance levels.[15]

Tolerances Ensure Food Safety

The tolerance is the residue level that triggers enforcement action. That is, if residues are found above the tolerance level, the commodity will be subject to seizure by the government. As discussed earlier, in setting the tolerance, the EPA must make a safety finding that the pesticide can be used by "reasonable certainty of no harm." To make this finding, the EPA considers the toxicity of the pesticide and its breakdown products, how much of the pesticide is applied and how often, and how much of the pesticide (that is, the residue) remains in or on food by the time it is marketed and prepared.

Some pesticides are exempt from the tolerance requirement. The EPA may grant exemptions in cases where the exemption is found to be safe. That is, the EPA must review toxicity and exposure data the same as for tolerance setting. There also must be a practical method for detecting and measuring levels of pesticide residues so regulatory officials can ensure that any residues are below the level found to be safe.

Other Agencies Involved

Several government agencies enforce EPA's pesticide tolerances in food. Besides the FDA's testing of domestically produced and imported food for residue limits, state enforcement agencies also check foods produced in this country.

The USDA tests meat and milk and, together with the FDA, has programs designed to develop statistically valid information on pesticide residues in foods. The USDA provides this information to the EPA to use in its risk assessment procedures for pesticides. If USDA workers detect violations of tolerances in their data collection program, they notify the FDA.

Numerous Scientific Studies Required

Pesticide manufacturers, or registrants, must submit a wide variety of scientific studies for review before the EPA will set a tolerance level. The data are designed to identify possible harmful effects the chemical could have on humans (its toxicity), the amount of the chemical (or breakdown products) likely to remain in or on food, and other possible sources of exposure to the pesticide (for example, through usage in homes or other places).

All of this information is used in the EPA's risk assessment process. Risk assessment includes consideration of the amounts and types of food people eat and how widely the pesticide is used (that is, how much of the crop is actually treated with the pesticide), as well as chemistry, toxicity, and exposure information. The EPA also uses data from the USDA on what foods people eat and the quantity they eat; this information is collected through the PDP.

The EPA Reassesses Old Tolerances

The EPA is reassessing all of the pesticide and other ingredient tolerances and exemptions that were in effect as of August 3, 1996, when the FQPA went into effect. This effort is designed to ensure that existing tolerances and exemptions meet safety standards set by the statute. The EPA is giving its highest priority to pesticides that appear to pose the greatest risk.

This reassessment is a huge task. More than 450 pesticides and other ingredients have tolerances or exemptions from tolerance requirements. There can be many tolerances associated with a given chemical, that is, a chemical might be used on various food crops, contributing to the complexity of the review.[16]

Monitoring Pesticide Residues

The FDA uses three approaches to monitor pesticides in domestically produced food shipped via interstate commerce and in imported food. They are regulatory monitoring, incidence/level monitoring, and data from the Total Diet Study.

Regulatory monitoring is directed toward enforcing tolerances in imported foods and in domestically produced foods shipped through interstate commerce. Under regulatory monitoring, the FDA samples individual lots of domestically produced and imported foods and analyzes them for pesticide residues. Emphasis is on the raw agricultural product, which is analyzed as the unwashed, whole (unpeeled), raw commodity. Processed foods also are included.

Domestic and imported food samples collected for analysis are classified as either "surveillance" or "compliance," that is, there is no prior knowledge or evidence that a specific food shipment contains illegal pesticide residues. Compliance samples are collected as a follow-up to the finding of an illegal residue or when there is other evidence of a pesticide problem. Compliance samples include follow-up samples from the same shipment as a violative surveillance sample, follow-up samples of additional products from the same grower or shipper, and audit samples from shipments presented for entry into the United States with a certificate of analysis (that is, shipments subject to detention without physical examination).

To analyze the large numbers of samples whose pesticide treatment histories are usually unknown, analytical methods capable of simultaneously determining a number of pesticides are used. These multiresidue methods (MRMs) can determine about half of the approximately 400 pesticides with EPA tolerances, as well as many others that have no tolerance. The most commonly used MRMs can also detect many metabolites, impurities, and alteration products of pesticides with and without tolerances.

Single-residue methods (SRMs) or selective MRMs may be used to determine pesticides not covered by an MRM. An SRM usually determines one pesticide; a selective MRM measures a relatively small number of chemically related pesticides. These types of methods are usually more resource-intensive per residue, and they may require at least as much time to perform as an MRM. They are much less cost-efficient than MRMs.

The lower limit of residue measurement in the FDA's determination of a specific pesticide is usually well below tolerance levels, which generally range from 0.1 to 50 parts per million (ppm). Residues present at 0.01 ppm and above are usually measurable; however, for individual pesticides, this limit may range from 0.005 to 1 ppm. In this database, the term "trace" is used to indicate residues detected, but at levels below the limit of measurement.

A complementary approach to regulatory monitoring, called incidence/level monitoring, has been used to increase the FDA's knowledge about particular pesticide/commodity combinations by analyzing certain foods to determine the presence and levels of selected pesticides. In 1995, a survey of triazines was begun; it was completed in 1997.

Statistically based monitoring surveys focusing on domestic and imported foods were initiated in 1992, and the results have been published. These surveys were initiated to determine whether FDA data acquired under regulatory monitoring are statistically representative of the overall residue situation for a particular pesticide, commodity, or place of origin. In the FDA's surveillance sampling for pesticide residues, sampling bias may be incurred by weighting sampling toward such factors as a commodity or a place of origin with a past history of violations or large volumes of production or import shipments. In addition, the total number of samples of a given commodity analyzed for a particular pesticide each year may not be sufficient to draw specific conclusions about the residue situation for the entire volume of that commodity in commerce. Therefore, the objective of these statistically based surveys is to determine whether violation rates, frequency of occurrence of residues, and residue levels obtained from such a sampling regimen differ from those obtained through the FDA's traditional surveillance approach.

Total Diet Study

The FDA's Total Diet Study (TDS) is designed to estimate dietary intakes of pesticide residues by men and women of various age groups, from infants to senior citizens. FDA personnel purchase foods from supermarkets or grocery stores four times per year, once from each of four geographical regions of the country. Each collection contains 261 food items (234 items prior to 1992) that are selected on the basis of information obtained through nationwide dietary surveys. The 261 foods are representative of more than 3,500 different foods in the national surveys; for example, apple pie represents all fruit pies and fruit pastries. Each of the four collections is a composite of similar foods purchased in three cities in that region. The foods are prepared and then analyzed for pesticide residues (as well as industrial chemicals, toxic elements, trace and macro elements, vitamin B6, and folic acid). The levels of pesticides found are used in conjunction with food consumption data to estimate the dietary intakes of pesticide residues.

Total Diet Studies are the primary sources of information on the levels of contaminants and nutrients in foods for human consumption. In addition, TDS results can be an indicator of environmental contamination by chemicals, such as Persistent Organic Pollutants (POPs), and can be used to assess the effectiveness of specific risk management measures. As the presence of toxic chemicals in our world and their potential presence in our food increase, it is increasingly important to assess human exposure to background concentrations of a large number of chemicals in the diet. The responsibility and obligation to make these assessments usually rests with national health authorities. Total Diet Studies are internationally recognized as the least expensive way to estimate the average dietary intakes of toxic and nutritional chemicals for a range of population groups.[17]

Organic Foods: Fewer Pesticide Residues

Do organically grown foods contain fewer residues of toxic pesticides than conventionally grown foods? The answer is an emphatic yes, according to a scientific study published in the peer-reviewed journal *Food Additives and Contaminants* in May 2002. In this detailed analysis of pesticide residues in foods, more than 94,000 conventional and organic food samples were studied by three organizations.

The USDA's data reported the following: 1) 73 percent of conventionally grown foods had one or more pesticide residues; 2) 23 percent of organically grown samples of the same crops had any residues; 3) more than 90 percent of conventionally grown apples, peaches, pears, strawberries, and celery had residues; and 4) conventionally grown crops were six times as likely as organic crops to contain multiple pesticide residues.

The California Department of Pesticide Regulation, using less-sensitive tests, found that: 1) 31 percent of conventionally grown samples had pesticide residues; 2) 6.5 percent of organic samples had pesticide residues; and 3) conventional samples had multiple residues nine times as often as organic samples.

Consumers Union showed the following: 1) 79 percent of conventionally grown samples had pesticide residues; 2) 27 percent of organically grown samples had pesticide residues; and 3) conventional samples had multiple kinds of residues ten times as often as organic samples.

Not only did fewer organic samples have any pesticide residues, but the level of pesticide residue found was consistently lower than the amounts found in conventional foods. The authors believe that residues found in organic foods could have been in the soil from previous pesticide use or could have come from pesticides sprayed on neighboring farms.[18]

The Reference Dose

How is pesticide safety determined for humans? The EPA establishes a reference dose (RfD) for each pesticide it approves for use. The RfD is the amount of a chemical that, if ingested over a lifetime, is not expected to cause any adverse health effects in any population subgroup. The RfD includes a ten- to 10,000-fold safety factor to protect humans over a lifetime, including infants, children, and other special populations. Using food consumption patterns and other data, the EPA estimates how much pesticide residue is likely to be consumed. If the RfD is exceeded the agency takes steps to limit the use of the pesticide.

Dose Response

Ironically, the extensive amount of data developed about a pesticide is often used against it by ignoring the dose response. For example, some acute toxicity studies, which are designed to include dosage levels high enough to produce deaths, are cited as proof of a chemical's dangers. Chronic effects seen at very high does in lifetime feeding studies are misinterpreted, according to the EPA, and considered as proof that no exposure to the chemical should be allowed, even though major improvements in analytical chemistry permit detection of the presence of chemicals at levels of parts per billion (ppb) or even parts per trillion (ppt).

We may hear that a certain chemical has been found in a food or beverage, and that the amount found is expressed in parts per million or parts per billion. Often, no information is provided to assist us in comprehending the meaning of these numbers. Frequently, this information neglects the issue of dose response, the key principle of toxicology, which simply stated is "the dose makes the poison." The concentration of a chemical in any substance is meaningless unless it is related to the toxicity of the chemical and the potential for exposure and absorption. Chemicals of low toxicities, such as table salt or ethyl alcohol, can be fatal if consumed in large amounts. Conversely, highly toxic materials may pose no hazard when exposure is minimal.

Maximum Residue Levels and Acceptable Daily Intake

For most of us, the primary exposure is what we eat and drink. Maximum Residue Levels (MRL) and Acceptable Daily Intake (ADI) are measures set by the federal government to assure us that human exposure to pesticides is limited. But there is no foolproof way to ensure a safe, universal ADI because of the diversity of food we eat and because some people are more vulnerable than others, especially young children and the malnourished. MRLs and ADIs also do not take into account the effects of combinations of pesticides or of pesticide breakdown products.

Very little research has been done to determine safe intake levels for the degradation chemicals of agricultural poisons. Consequently, there are virtually no safety levels to determine the ADI of toxic breakdown chemicals that contaminate our food.

"Legal" Does Not Equal "Safe"

Fortunately, most pesticide residues detected in or on foods are within the legal limits established by the EPA. Does that mean the residues are "safe"? Not at all. Why is food safety an ephemeral concept? There are several explanations. First, tolerances were set, often many years ago, by estimating maximum residues on foods after pesticide treatments. Most tolerances for older pesticides were set with little or no assessment of risks to human health, but were intended as a basis for monitoring to ensure that pesticide uses were in accord with label directions. Also, since tolerances are the only legal limits for residues in domestic or imported foods that the government sets, federal law and many scientific advisory bodies have directed the EPA to use them to keep dietary pesticide exposure to within "safe" limits. However, U.S. pesticide legislation also required the EPA to balance the goals of permitting pesticide use and protecting human health. Consequently, many tolerances were set at levels higher than safety concerns alone would require.

The FQPA requires that pesticide exposures be "safe," including safe for infants, young children, and other sensitive subpopulations. The law places primary emphasis on safety, leaving the EPA much less room to trade off health protection against the economic benefits of pesticide use.

What is meant by "safe?" The FQPA, as we have seen, defines it as "reasonable certainty of no harm" to public health. In practice, scientists seeking to define safe levels of exposure to chemicals generally rely on toxicity data from animal tests and incorporate a "safety margin" to take into account the scientific uncertainties involved. Using this approach, the EPA, with peer review by the outside scientific community, establishes a "reference dose," or RfD for each pesticide. The RfD, measured in milligrams of chemical per kilogram of body weight per day, defines a dose level that is thought to be without appreciable risk to human health, that is, a "safe" intake.

To assess whether legal tolerances establish safe levels of exposure, the following computation is instructive:

Compare the EPA's current tolerances for pesticides found in the USDA's PDP tests to the RfD's tests for those pesticides. To facilitate comparisons, a reference concentration (RfC) is calculated. The RfC is the residue concentration that gives a twenty-kg child (forty-four pounds, about average for a five-year-old) who eats 100 grams (about 3.5 ounces) of a food item a reference dose of the pesticide. A different body weight or serving size of food would result in a different RfC.

The RfC can then be compared with the tolerance. If the tolerance is greater than the RfC, it means *a twenty-kg child who ate 100 grams of this food would exceed the RfD for that pesticide on that day from the single serving of a single food* (italics mine). Now, this does not mean the child's health would automatically or immediately be harmed; the RfD includes a "safety margin." But it does mean that the safety margin between actual doses and those known to be harmful *is narrower than it should be* (italics mine). The definition of "safe" for pesticide residues ostensibly includes *ensuring an adequate safety margin* (italics mine). By definition, then, *tolerances that permit exposures above the RfD cannot be considered "safe" limits*. Certainly, they are not "safe enough."

Unfortunately, then, most of the tolerances for these pesticides on these foods exceed the RfD, often by a wide margin. For example, tolerances for methyl parathion would allow 250 times the safe dose on all six foods on which this insecticide was detected by the PDP. Likewise, tolerances for acephate and dicofol allow up to forty-two times the RfC, those for chlorpyrifos allow up to twenty-five times the RfC, and most tolerances for dimethoate/omethoate permit twenty times the RfC. The "action level" for dieldrin (there are no tolerances for banned pesticides such as dieldrin) is ten times greater than the RfC.

These comparisons make it clear that there is a big difference between "legal" residues and "safe" residues. Consumers can take little comfort from the fact that most residues are within the tolerances set by the EPA, because many tolerances currently on the books *permit unsafe exposures* (italics mine).[19] What should the EPA do to overcome the tendency to assume that computations from testing done by pesticide manufacturers invariably results in both "legal" and "safe" pesticide residues on food? There is no doubt that as scientific understanding of potential cumulative and aggregate effects advances, it is certain that additional and more precise data will be required for EPA decisions (we cannot assume that the EPA will recognize this), along with more information on subpopulation exposure and risk. In most cases the EPA will be able to use existing FIFRA authority to require this data. Additional data will hopefully enhance the scientific basis and effectiveness of pesticide regulations.[20]

It may seem like an incidental detail, but we cannot overlook an important limitation on the accuracy of tolerance setting: the procedures used to set tolerances in the past neglect the fact that per capita consumption of some fruits and vegetables has risen, causing potential understatement of residue intake. For example, average consumption of fresh honeydew melons, broccoli, and tomatoes increased from 1.1, 1.1, and 12.6 pounds per person, respectively, in 1976, to 2.4, 3.3, and 16.8 pounds in 1987. Also, over the past two decades, the average American consumed 20 percent more fruits, vegetables, and grain products than in prior years. Some of the pesticides used on these commodities were registered before consumption increased.[21]

Required Tests

The EPA requires a battery of toxicity tests in laboratory animals to determine a pesticide's potential for causing adverse health effects, such as cancer, birth defects, and problems with the nervous system or other organs. Tests are conducted for both short-term (acute) and long-term (chronic) toxicity. The RfDs are computed following administration of different doses of a pesticide to determine the level at which no adverse effects occur. Calculation of an uncertainty or "safety" factor (usually 100) accounts for the uncertainty of extrapolating from laboratory animals to humans and for individual human differences in sensitivity. For cancer risks, the EPA evaluates multiyear tests of laboratory animals to estimate levels unlikely to pose more than a negligible risk.

Several of the types of studies that the EPA can require are designed specifically to assess risks to infants and children. These include developmental toxicity studies, which examine risks to developing fetuses from exposure to pesticides during pregnancy; developmental neurotoxicity studies, which examine the risks to the developing nervous system; and two-generation reproduction studies, which provide information about the effects on the health of both an individual and his or her offspring due to pesticide exposure.[22]

Toxicity Testing Deficiencies

The toxicity testing that the EPA requires of pesticide manufacturers, which is largely performed in laboratory animals and used in setting tolerances, is often inadequate to protect children. EPA guidelines used by manufacturers to complete these tests reveal data gaps that fall into several categories.

Most toxicity testing for food-use pesticides uses only adult animals. Of the two tests required for food-use pesticides that actually do expose developing animals, one fails to continue dosing the animal after birth, when many organ systems are still maturing. How can pesticide tolerances based on these tests carry a reasonable certainty of no harm to infants and children? In fact, specific testing for toxicity to the immune system has been required for only two chemical pesticides, and developmental neurotoxicity testing has only been completed for six pesticides. Children depend on healthy brains and nervous and immune systems to become educated, productive adults.

In compliance with the FQPA, the EPA is still developing guidelines for testing a chemical's potential for disrupting normal functions of the endocrine (hormone) system. Normal development of the fetus, infant, and child depends on the timely release of low levels of various hormones and their effects on different organs. Until guidelines and testing are implemented, a pesticide's untested potential for endocrine disruption should be reflected in the tolerance for that chemical.[23]

Pesticides on Fruits and Vegetables

A detailed report revealed unsafe levels of pesticide residues on certain fresh fruits and vegetables, including many that are grown in the United States. Produce was tested and scored based on how many samples contained pesticides, the average amount detected, and the toxicity of the particular pesticides found. The test data was from USDA testing, and the foods were prepared as they would be at home (for example, bananas and oranges were peeled). Each sample was a composite of about five pounds of produce. A score greater than 100 is cause for concern.

Here are the worst:

Type of Produce and Score

Peaches, domestically grown (North America): 4,848

Peaches from Chile: 471

Winter squash, domestically grown: 1,708

Apples, domestically grown: 550

Pears, domestically grown: 435

Pears from Mexico: 415

Spinach, domestically grown: 349

Spinach from Mexico: 256

Grapes, domestically grown: 228

Grapes from Chile: 339

Celery, domestically grown: 255

Green beans, domestically grown: 222

Surprisingly, bananas, which used to be heavily dosed with pesticides, scored only four points. One chemical, methyl parathion, accounted for more than 90 percent of the total toxicity load of peaches, apples, pears, green beans, and peas. The high toxicity values for winter squash from the United States were almost entirely due to residues of dieldrin, a very toxic, carcinogenic insecticide that was banned twenty-five years ago but still persists in some agricultural soils.

Illegal insecticides found on produce were not the result of excessive residues, but rather were due to low levels of chemicals that are persistent residues in soils, or to wind dispersal of pesticides applied legally to nearby fields. But the data revealed widespread illegal use of several insecticides on spinach grown in both the United States and Mexico.

It is commonly thought that foods imported from Mexico and South America are more contaminated than food grown in the United States. However, eleven of the twelve most contaminated foods in the study were U.S. grown. The worst offenders included Chilean grapes, Canadian and Mexican carrots, Mexican broccoli and tomatoes, Argentine and Hungarian apple juice, and Brazilian orange juice. Samples of fresh peaches, fresh and frozen winter squash, fresh green beans, apples, and pears grown in the United States were more contaminated than imports.

Least Contaminated: Consistently Clean

The study found that the vegetables least likely to have pesticides on them are sweet corn, avocado, cauliflower, asparagus, onions, peas, and broccoli.

Nearly three-quarters (73 percent) of the pea and broccoli samples had no detectable pesticides. There were no detectable residues on 90 percent or more of the samples of the other vegetables on the least-contaminated list.

The study revealed that multiple pesticide residues were extremely rare on the least-contaminated vegetables. Broccoli had the highest likelihood of contamination, with a 2.6 percent chance of containing more than one pesticide. Avocados and corn both had the lowest chances, with no samples containing more than one pesticide.

The greatest number of pesticides detected on a single sample of any of these low-pesticide vegetables were three compared to ten found on spinach, the most-contaminated crop with the most residues.

Broccoli and onions both had the most pesticides found on a single vegetable crop at up to seventeen. But they had far fewer than the most-contaminated vegetable, sweet bell peppers, on which thiry-nine were found.

The five fruits least likely to have pesticide residues on them are pineapples, mangoes, bananas, kiwi, and papaya. Fewer than 10 percent of pineapple and mango samples had detectable pesticides on them and less than 1 percent of samples had more than one pesticide residue. Though 53 percent of bananas had detectable pesticides, multiple residues were rare, with only 4.7 percent of samples containing more than one residue. Kiwi and papaya had residues on 23.6 percent and 21.7 percent of samples, respectively, and multiple pesticide residues of just 10.4 percent and 5.6 percent of samples, respectively.[24]

There are some pesticide residues detected in food for which no tolerances have been set. This could be due to the following reasons: 1) the pesticide may be recognized by the EPA as safe and not requiring tolerances, 2) the pesticide may not be registered for agricultural use in the United States, or, more often, 3) the pesticide residue is on a crop different from the one for which it has been approved. For some residues of this type, the FDA may enforce international action levels. Alternatively, the pesticide residues may be illegal in food in the United States, so the food can be seized and destroyed.[25]

Harmful Breakdown Products of Pesticides

In some cases, when a pesticide is known to have harmful breakdown products, a tolerance limit may be set for the total amount of "parent" pesticide and breakdown products that may be present in or on food. In such cases, foods may be tested for all these residues. For example, tolerances have been established for the total amount of the pesticide endosulfan and its breakdown product endosulfan sulfate.

Permethrin

The FDA's monitoring program routinely finds the pesticide permethrin on food. In 1996, it was the thirteenth most commonly detected pesticide. Similar results were found in the USDA's monitoring of fourteen fruits and vegetables. Permethrin was the tenth most frequently detected pesticide. It was found on spinach in 60 percent of the samples tested and on 11 percent of the tomato samples tested. Permethrin was also frequently detected on celery and lettuce. It has even been found in baby food. FDA monitoring discovered it in 12 percent of samples tested. The Environmental Working Group discovered that permethrin was the most commonly detected pesticide in peach baby food, in some 44 percent of samples tested. It was also found in plums in 11 percent of samples.[26]

Daily Pesticide Exposures

Through their diets, U.S. consumers can be exposed up to seventy times each day to residues from persistent organic pollutants (POPs). The use of POPs is not allowed in organic agriculture. The top ten POP-contaminated food items, in alphabetical order, are butter, cantaloupe, cucumbers/pickles, meatloaf, peanuts, popcorn, radishes, spinach, summer squash, and winter squash. The two most pervasive POPs in food are dieldrin and DDE, a compound produced when DDT degrades.[27]

More than one million children between the ages of one and five ingest at least fifteen pesticides every day from fruits and vegetables. More than 600,000 of these children eat a quantity of organophosphate insecticide that the federal government considers unsafe, and 61,000 eat doses that exceed benchmark levels by a factor of ten or more.[28]

See Tables 3.1–3.4 for more details on pesticide dosages.

The Nondetect Factor

As mentioned earlier, in some cases, a portion of the measurements of the levels of pesticide residue present on food shows no detection of residues. These "nondetects" (NDs) do not necessarily mean that the pesticide is not present, but simply that any amount of pesticide present is below the level that could be detected or reliably measured using a particular analytical method.

Table 3.1

More than 600,000 Children under Age Six Get an Unsafe Dose of Neurotoxic Pesticides in Food Each Day

Age	Est. no. of children exceeding EPA "safe" dose/day	Percent of population	Est. no. of children exceeding 10 times EPA "safe" dose/day
1	137,200	3.4	13,500
2	131,400	3.3	13,500
3	130,000	3.2	13,900
4	104,300	2.6	9,700
5	107,600	2.7	10,400
Total	**610,500**	**3.1**	**61,000**

Source: Compiled from USDA food consumption data, 1989–1996; USDA and FDA pesticide residue data, 1991–1997; and EPA data, 1998a and 1998b.

Table 3.2

Apples and Apple Products Account for More than Half the Unsafe Organophosphate Insecticide Exposure for Children under Age Six

Food	Est. no. of children exceeding EPA "safe" dose/day from individual foods					
	1-year-olds	2-year-olds	3-year-olds	4-year-olds	5-year-olds	Total
Apples	32,430	51,050	54,720	48,370	48,380	234,950
Peaches	15,450	11,670	16,220	9,570	11,430	64,340
Green beans	12,320	9,870	13,360	10,830	10,550	56,930
Applesauce	15,440	9,610	10,620	7,430	12,360	55,460
Apple juice	18,920	14,350	9,110	5,470	2,830	50,680
Grapes	10,040	11,850	10,110	8,440	8,610	49,050
Pears	4,410	5,470	3,040	3,380	2,960	19,260
Nectarines	1,610	1,210	2,080	1,670	2,970	9,540
Tomatoes	1,460	500	640	780	780	4,160
Raisins	340	1,550	280	470	0	2,640
Strawberries	330	500	390	440	430	2,090
Plums	380	450	360	620	260	2,070
Bell peppers	40	440	210	410	460	1,560
Spinach	210	350	310	110	320	1,300
Tangerines	280	90	180	210	410	1,170

Source: Compiled from USDA food consumption data, 1989–1996; USDA and FDA pesticide residue data, 1991–1997; and EPA data, 1998a and 1998b.

Table 3.3
Parents Can Reduce Health Risks to Their Children by Feeding Them Fruits and Vegetables with Consistently Low Pesticide Residues

Most-contaminated foods		Least-contaminated foods	
Rank	Food	Rank	Food
I	Apples	I	Corn
2	Spinach	2	Cauliflower
3	Peaches	3	Sweet Peas
4	Pears	4	Asparagus
5	Strawberries	5	Broccoli
6	Grapes (Chile)	6	Pineapples
7	Potatoes	7	Onions
8	Red raspberries	8	Bananas
9	Celery	9	Watermelons
10	Green beans	10	Cherries (Chile)

Source: Compiled from USDA and FDA pesticide residue data, 1992–1997.

Table 3.4
Older, Highly Toxic Pesticides Continue to Dominate the Market

Pesticide	Product first registered	Est. total use in lbs. (1995)	Health effects
Altrazine	1959	70,500,000	Carcinogen, hormone disrupter, tap water contaminant
Metolachlor	1977	61,500,000	Carcinogen, hormone disrupter tap water contaminant
2,4-D	1948	53,000,000	Neurotoxicant, hormone disrupter
Metam sodium	1955	51,500,000	Carcinogen, teratogen
Methyl bromide	1947	48,500,000	Lethal neurotoxicant, teratogen, depletes ozone layer
Glyphosate	1974	43,000,000	Neurotoxicant
Dichloropropene	1960	40,500,000	Carcinogen
Cyanazine	1971	26,500,000	Teratogen, carcinogen, tap water contaminant
Pendimethalim	1975	25,500,000	Carcinogen
Trifluralin	1963	25,500,000	Carcinogen, hormone disrupter
Chlorpyrifos	1965	25,000,000	Developmental neurotoxicant

Source: EPA pesticide industry sales and usage: 1994 and 1996 market estimates, August 1997; health information from EPA Office of Pesticide Programs data.

The primary science policy issue concerning NDs is what value the EPA's Office of Pesticide Programs (OPP) should assign to them when estimating exposure and risk from pesticides in food. The reason this is an important issue stems from the requirements that the FQPA imposes on the EPA. The OPP's claimed objective is to make exposure and risk assessment as accurate and realistic as possible while not underestimating exposure or risk, so that all people, including infants and children, are fully protected.[29]

Risk of Exposure Calculation

Potential exposure to a chemical in a specific food is assessed by multiplying the residue concentrations in food by the amount of food consumed. Potential dietary exposure to a chemical is assessed by tabulating residue intakes from all foods.

Different assumptions regarding residue concentrations in food may be used to assess exposure. A worst-case exposure scenario may be calculated using tolerance levels for pesticides in food. This exposure assessment is the theoretical maximum residue contribution. Exposure may also be calculated using anticipated residue levels.

Pesticide Inerts Treatment

The EPA is required to set tolerances or grant exemptions based on the FQPA safety standard for all ingredients in a pesticide product for use on food. The law stated that inert ingredients used on food are "pesticide chemicals" on the same basis as active ingredients. Inerts are approved under the same safety standard of "reasonable certainty of no harm" from dietary exposures and all other exposures, where there is reliable information, for all people, including infants and children. Inerts are now referred to as "other ingredients."

For that matter, approvals have been very slow and practically nonexistent for many food-use chemicals. For these so-called other ingredients the EPA has set one tolerance and exempted seventeen from tolerance requirements based on the FQPA's standard.[30]

Storage and Processing Impact

There is a large gap between consumer and scientific perceptions on the risks that pesticide residues in food pose to human health relative to other dietary risks. One cause of this misconception has been the emphasis placed on "worst-case" evaluations and extrapolations of available data, for example, assuming that all crops are treated with pesticides and that the resulting residues in food are at maximum permitted levels.[31]

Controls on pesticide residues in crops are generally based on MRLs that are set using field trial data to arrive at the highest expected pesticide residue levels. Primary residue studies on food crops are mainly carried out on samples that are deeply frozen prior to analysis and that receive minimal post-harvest handling, except possibly minor trimming. Although MRLs are a credible and useful means of enforcing acceptable pesticide use, they are inadequate as a guide to human health risks from residues. Total Diet Studies have consistently shown that using MRLs as a basis for calculating human dietary consumption of pesticides overestimates actual intakes by one to three orders of magnitude. In addition, processing food for consumption can lower perceived pesticide exposure and intake. A recommended approach to evaluating risks from pesticide residues in food includes allowances for losses in processing.

Processing treatments such as washing, peeling, canning, or cooking are important factors leading to the reduction of pesticide residues left on crops. Most foods are treated in some fashion before they are consumed. Processing can often substantially reduce residue levels on or in food that has been treated with pesticides. For example, a study tracking chlorothalonil on crops from field to table showed that normal handling and processing of fresh cabbage, celery, cucumbers, and tomatoes led to large reductions in residue levels.[32] The actual exposure of U.S. consumers to chlorothalonil through diet was calculated at only 2 percent of the maximum theoretical level estimated from MRLs. However, in some special cases, processing can cause more toxic by-products or metabolites to form. Processing can also result in residues being redistributed or concentrated in various parts of food.

Regulatory authorities are increasingly interested in such data. Studies into the effects of storage and some commercial processing techniques on residues in food are a part of the registration process for pesticides in many countries. Data on processing are considered necessary to reassure consumers as to actual versus hypothetical exposure to food residues.

Organophosphates

Organophosphates pesticides (OPs) frequently are applied to many of the foods important in children's diets, and certain OP residues can be detected in small quantities. When exposure to OPs is sufficiently high, as previously indicated, they interfere with the proper functioning of the nervous system. There are approximately forty OPs, and as a group they account for approximately half of the insecticide use in the United States. The majority of OP use is on food crops, including corn, fruits, vegetables, and nuts. In addition, OPs often have been used in and around the home. Examples of OP pesticides include chlorpyrifos, azinphos methyl, methyl parathion, and phosmet.

Between 1994 and 2001, from 19 percent to 29 percent of food samples had detectable OP residues. The highest detection rates were observed during 1996 and 1997, and the lowest rate was seen in 2001. Between 1993 and 2001, the amount of OP pesticides used on foods most frequently consumed by children declined by 44 percent from 25 million pounds to 14 million pounds. In 1999 and 2000, the EPA imposed new restrictions on the use of the OPs azinphos methyl, chlorpyrifos, and methyl parathion on certain food crops and around the home, due largely to concern about potential exposure to children.

Thirty-four OPs were sampled in each of these years. This measure is a surrogate for children's exposure to pesticides in foods. If the frequency of detectable levels of pesticides in foods decreases, it is likely that exposures will be reduced. However, this measure does not account for many additional factors that affect exposure risk to children. For example, some OPs pose greater risks to children than others, and residues on some goods may pose greater risks than residues on other foods due to differences in the amounts consumed. In addition, year-to-year changes in the percentages of samples with detectable residues may be affected by the selection of foods that are sampled each year. There is growing evidence that OPs are toxic to the developing brain as well as the nervous system at very low levels of exposure. One carcinogenic OP pesticide was taken off the market in 1974, but it is still found in farming soil at such high levels that there is a 77 percent chance that a child will get too much in a single serving of winter squash.[33]

Overexposed: OPs in Children's Food

Every day, nine out of ten American children between the ages of six months and five years are exposed to combinations of thirteen different neurotoxic insecticides in the foods they eat. While the amounts consumed rarely cause acute illness, OPs pose a serious threat to children, who are rapidly growing and extremely vulnerable to injury during fetal development, infancy, and early childhood.

The following estimates are based on the most recent EPA data available on children's eating patterns, pesticides in food, and the toxicity of OP insecticides:

Every day, more than one million children ages five and under (one out of twenty) eat an unsafe dose of OP insecticides. One hundred thousand of these children are exposed to more than the EPA's safe dose, the reference dose, by a factor of ten or more. For infants six to twelve months of age, commercial baby food is the dominant source of unsafe levels of OP insecticides. Baby food, apple juice, pears, applesauce, and peaches expose about 77,000 infants each day to unsafe levels of OP insecticides. This estimate likely understates the number of children at risk because the analysis does not include residential and other exposures, which can be substantial. In addition, because the EPA's estimates of a safe daily dose are based on studies on adult animals or humans, they almost never include additional factors that would shelter the young from the toxic effects of OPs.

This data analysis also identified foods that expose young children to toxic doses of OPs, finding:

One out of every four times a child age five or under eats a peach, he or she is exposed to an unsafe level of OP insecticides. Thirteen percent of apples, 7.5 percent of pears, and 5 percent of grapes in the U.S. food supply expose the average child to unsafe levels of OP insecticides.

A small but worrisome percentage of these fruits—1.5 to 2 percent of apples, grapes, and pears, and 15 percent of peaches—are so contaminated that the average twenty-five-pound one-year-old eating just two grapes or three bites of an apple, pear, or peach (ten grams of each fruit) will consume more than the EPA's safe daily adult dose of OPs.

Because they are more heavily consumed, apples, peaches, applesauce, popcorn, grapes, corn chips, and apple juice expose the most children ages six months through five years to unsafe levels of OPs. Just over half of children who eat an unsafe level of OPs each day, 575,000, receive this unsafe dose from apple products alone. Many of these exposures exceed safe levels by wide margins. OPs on apples, peaches, grapes, pear baby food, and pears cause 85,000 children each day to consume more than the federal safety standard by a factor of ten or more.[34]

OP Residue "Hot Spots"

The FQPA requires the use of an additional safety factor to account for the fact that some pesticides pose greater toxicological risks to fetuses and young children than to adults. According to the act, an added safety factor should also be imposed in cases where there are gaps or uncertainties in exposure estimates. Also, common high-end residues in several children's foods can expose a child to more than what would be regarded as acceptable, even for residues in a single food consumed during a given day.

There are, in fact, several dozen crop-pesticide combinations that will periodically result in residues high enough to put some children over the EPA's daily acceptable OP exposure levels. The EPA has no control over and only modest potential to pre-dict where and when such pesticide residue "hot spots" will materialize. The only thing that is predictable is that some will occur each year across the United States and around the globe, driven by unusually intense pest pressure and/or the collapse of effective pest management systems. This is often associated with the emergence of new and/or resistant strains of pests.

The result will be a sudden increase of relatively high OP residues in the American diet stemming from crop-pesticide combinations that have never been seen before. These will contribute markedly to exposure risk. These OP hot spots will make attainment of the basic FQPA safety standard fleeting, and, over the long term, nearly impossible, unless the EPA takes actions across eighty to 120 OP-crop uses to reduce their likelihoods and severity.

In the revised organophosphate-cumulative risk assessment (OP-CRA), the use of dimethoate on grapes is an example of a residue-driven hot spot. Dimethoate accounted for almost one-half of the total OP risk among one- to two-year-old chil-dren. While this accounts for a major risk factor among all food-OP combinations, it would be wrong to conclude that the problems posed by dietary OP residues can be solved by phasing out this and a half-dozen other uses that, in this particular OP-CRA, account for the lion's share of risks. This is because the revised OP-CRA results are a snapshot of a very complex landscape that can and does change dramatically with the seasons and as pest pressure varies across different crops.

It is important to point out a strong downward bias in the revised OP-CRA results. The risk levels projected take into account OP risk mitigation measures to date. But the results are also based on an implicit assumption that farmers who cannot use a banned OP will not switch to another OP. Clearly, some shift in acreage from canceled OPs to acceptable ones is occurring and will continue. Given the lag between the use of pesticides and the collection of USDA PDP residue data, the results of the revised OP-CRA do not reflect the greater frequency of residues nor higher levels for OPs that will be used more intensively in the future than they were used in the past.

There is another source of significant uncertainty in projecting residue hot spots. Although the EPA imposes risk mitigation measures in the United States via changes in product label, the agency's ability to project changes in residue profiles in imported foods is often limited. Since imported foods and vegetables make up a large and growing portion of the U.S. market, the EPA must rely heavily on the one regulatory tool that can and will directly impact the share of OP risk accounted for by imports—tolerance reductions.[35]

Organochlorine Residues

Organochlorines are some of the chemicals found most often in hundreds of tests worldwide of human body tissue such as blood, fatty tissue, and breast milk. Because of their chemical structure, organochlorines break down slowly, build up in fatty tissue, and remain in our bodies for a long time. Pesticide residues on food are a major source of organochlorine exposure. An investigation of organochlorine residues in the U.S. food supply found that even those chemicals that have been banned for decades still are found consistently in food samples tested by the FDA. This can be explained in part by the long life spans of many organochlorines in the environment. Dieldrin and the breakdown products of DDT, for example, can remain in soil for decades. In addition, other countries continue to use organochlorine pesticides on food that is imported into the United States, and residues can be transported via wind and water currents.

Organochlorines contribute to many acute and chronic illnesses. Symptoms of acute poisoning can include tremors, headache, dermal irritation, respiratory problems, dizziness, nausea, and seizures. Organochlorines are also associated with many chronic diseases. Studies have found a correlation between organochlorine exposure and various types of cancer, neurological damage (several organochlorines are known neurotoxins), Parkinson's disease, birth defects, respiratory illness, and immune dysfunction. Many organochlorines are known or suspected hormone disrupters, and there is evidence that even extremely low levels of exposure in the womb can cause irreversible damage to the reproductive and immune systems of the developing fetus.

Are Organochlorines Regulated?

The United States and other countries have banned many organochlorines because of concerns about environmental impacts and human health effects. In addition to DDT, the United States has banned aldrin, dieldrin, arochlor, chlordane, heptachlor, mirex, hexachlorobenzene, oxychlordane, toxaphene, and others. However, several organochlorines are still registered for use, including lindane, endosulfan, methoxychlor, dicofol, and pentachlorophenol.

Some organochlorines have been targeted for global elimination under the recently signed Stockholm Convention on Persistent Organic Pollutants. The treaty is an international effort to phase out harmful chemicals that persist in the environment and can be transported around the world. Many organochlorines fall into this category. The initial list of twelve chemicals targeted by the treaty included nine organochlorine pesticides that have already been banned in the United States. The United States has not yet ratified the Stockholm Convention largely due to resistance from the current presidential administration regarding the process of listing new POPs that are still in use in the United States.[36]

The EPA can manage the risks of currently used pesticides by setting strict tolerance limits to keep residues within the range required by the FQPA. The dietary risk contributions of dicofol and endosulfan, for example, could be managed this way. But for a banned chemical, such as dieldrin or heptachlor, EPA tolerances are already set at zero. Unavoidable residues caused by environmental contamination are legal and are governed by "action levels" set by the FDA. An action level defines a level of contamination that may render a food injurious and warrants keeping it off the market.

Current action levels for banned organochlorine insecticides are relatively high; the action level for dieldrin is 0.1 ppm. High action levels for the banned organochlorines sanction serious residue problems, such as those observed in winter squash. As long as it remains legal, squash growers will continue to sell products that contain significant dieldrin and heptachlor residues. If these action levels were lowered, say to 0.01 ppm, growers would have an incentive to seek out uncontaminated land for food crops that absorb organochlorines as effectively as squash does. The FDA depends on the EPA for risk assessments on pesticides. To provide a basis for setting more health-protective action levels for the banned organochlorine pesticides, the two agencies need to work together. Under the FQPA, ensuring a wider safety margin for these residues should be a high priority.[37]

Status of Carbamate Exposures

Despite the clear advice from its own scientific advisory panel and innumerable other expert bodies, the EPA has stuck by its decision to exclude other cholinesterase-inhibiting pesticides from the revised and final OP-CRA. A preponderance of scientific evidence supports the inclusion of the carbamate insecticides in any cholinesterase inhibition-based cumulative risk assessment.

Excluding carbamates has consequences. Farmers use insecticides from these two chemical families interchangeably, often for resistance management purposes. In most crops, carbamates leave residues about as frequently as OPs and often at comparable levels.

Unless the EPA includes both OPs and carbamates in future cumulative risk assessments, and imposes comparable risk mitigation measures across all OPs and carbamates registered for use on a particular crop, the adverse health consequences the EPA is trying to prevent—cholinesterase inhibition and developmental neurotoxicity—will persist.[38]

The "Channels-of-Trade" Policy

Even when the EPA negotiates a partial or total pesticide ban, it can be years before foods containing the chemical clear store shelves. Written into the FQPA of 1996 is a "channels-of-trade" provision, which assures that, in the event of a ban, producers are not immediately responsible for removing the pesticide from circulation. For example, it may still be in interstate commerce by the time the revocation or new lower tolerance level takes effect. Such a food could be found by the FDA to contain a residue of a revoked pesticide or an amount of residue that exceeds the new lower tolerance. The FDA would normally deem such a food to be in violation of the law by virtue of it bearing an illegal pesticide residue. The food would be subject to an FDA enforcement action as an "adulterated" food under the FFDCA. However, the channels-of-trade provision provides an exception to such a finding by the FDA provided that certain criteria are met.

The FDA gives firms the opportunity to demonstrate the last date that the FDA anticipates that food made from lawfully treated commodities will remain on the market. For certain processed foods, that is, frozen, dried, and canned foods, this date is generally four years from the time the treated crop is harvested. It should be noted that this opportunity is not granted under the FFDCA for food bearing pesticide residues that are not potentially subject to the channels-of-trade provision, such as when the residue of a revoked pesticide in a food exceeds the prior tolerance for the food or when a pesticide residue is found in a food for which no tolerance exists.[39]

Regulation of OP Residues

In agriculture, OPs such as methyl parathion and malathion are broadly effective insecticides, killing boll weevils or fruit flies, for example. Various OPs are used on fruit trees, vegetables, ornamental plants, cotton, corn, soybeans, rice, and wheat, and for mosquito control. Some are acutely toxic, others much less so, but because they exert dangerous health effects in the same way by interfering with the proper functioning of the nervous system, they were the first pesticides considered as a group for tolerance reassessment by the EPA. In 1996, 1,691 tolerances were assessed for OP residues on crops. By August 2, 2002, the EPA had assessed 1,127 tolerances (about 67 percent), and revoked 703 OP tolerances.

Growers and pesticide manufacturers are concerned about the FQPA's impact on future availability of widely used OP pesticides. The EPA already has canceled methyl parathion registrations for all fruit uses. In June 2000, the EPA and the manufacturer of the OP chlorpyrifos (Dursban) agreed to eliminate nearly all household uses and reduce residues on several foods regularly eaten by children. In December 2000, the EPA announced a plan to phase out all home uses of diazinon, another widely used

OP pesticide. In 2001, the EPA decided to cancel, phase out, or continue under time-limited registrations the crop uses of azinphos methyl and phosmet. In addition, the EPA and registrants will voluntarily ban certain uses of propargite. The goal of these and other regulatory actions against individual OPs is to reduce cumulative OP risk to a safe level.

The EPA released a preliminary cumulative risk assessment for OPs in December 2001 and revised cumulative OP risk assessment in June 2002. However, final OP cumulative risk assessment is not yet complete. As the EPA collects more data, revised cumulative OP risk assessment might produce lower risk estimates, indicating that pesticide residue levels are reasonably certain to be safe (as farmers and pesticide manufacturers contend). On the other hand, data might support the view of public health advocacy groups that indicates that, despite regulatory actions, children are exposed to unsafe levels of OPs on pears, apples, grapes, and peaches, risking damage to developing brains and nervous systems.[40]

The "Body Burden"

Just as we don't tend to think about the invisible, we seldom consider the accumulation of years of pesticide exposure in our bodies. While some sources believe the benefits of pesticides to human nutrition outweigh any potential health risks, research suggests otherwise.

When veteran journalist Bill Moyers had his blood and urine tested as part of a Mount Sinai School of Medicine study of pollutant loads in the human body, eighty-four distinct chemicals were found, including some that had been banned more than a quarter of a century earlier. This chemical "body burden," as it is medically known, is even more insidious for children, whose developing bodies and brains are more vulnerable than those of adults. A 2005 *Journal of the American Medical Association* study found that the incidence of illness linked to pesticide use in and around U.S. schools is rising. Organophosphates, the same class of insecticides detected in Moyers's blood, were most frequently responsible for these poisonings.[41]

Pesticides and the Enzyme PON-1

Pesticide regulation might not be strict enough to protect newborns and infants. A 2006 University of California study of 130 mothers and their children in California's Central Valley, a large agricultural area, revealed that a natural enzyme that breaks down toxicants (including commonly used pesticides) varies to such a degree that some young people may be virtually defenseless against many chemicals. This enzyme, PON-1, is one of hundreds of important enzymes that control the body's metabolism. Other studies have shown that PON-1 is linked to protection against neurodegenerative or cardiovascular diseases.

The study was designed to examine the protective levels of PON-1 against OPs. In pesticides, OPs attack the nervous systems of insects. Two common OPs, diazinon and chlorpyrifos, were widely used before they were restricted from most household uses by California and the EPA in 2002 (chlorpyrifos was previously a key ingredient in the pesticide Raid). An analysis of PON-1 in newborns found that those with lower levels of the enzyme may be twenty-six times more susceptible to diazinon exposure than those with the highest level of the enzyme. The former may be sixty times more susceptible than adults with the highest enzyme levels.

With chlorpyrifos, some of the newborns may be fifty times more susceptible than newborns with high enzyme levels and 130 to 164 times more susceptible than some of the adults. The enzyme typically reaches adult levels by the time children reach two years of age. The two pesticides are still used on cropland under the brand names Dursban and Lorsban. Chlorpyrifos was banned in households largely because of its hazards to children but it is still widely used in farm applications.[42]

Malathion and Methyl Parathion

Based on the FDA's residue analyses, malathion has been the most commonly detected pesticide in food products. Malathion residues were in 18 percent of 936 food items tested, indicative of its widespread use in many crops. It is also commonly found in animal feeds.[43] In one greenhouse study, malathion applied at recommended rates was easily detected on plant surfaces up to nine weeks after spraying.[44]

Another pesticide, methyl parathion, is so toxic that a five-year-old cannot eat an apple with any detectable methyl parathion on it without being exposed to an unsafe dose. Some apples and peaches are so contaminated that just two bites (four to seven grams) are unsafe for children under age six. Any other organophosphate exposure in food, at home, at school, in water, or in the air will push that child's daily exposure to these compounds further into the zone that the EPA already considers unsafe. In August 1999, the EPA accepted voluntary cancellation of many of the most significant food crop uses of methyl parathion. Reducing these crop uses considerably reduced risks to children through food, as well as risks to workers and the environment.

The EPA's risk assessment showed that methyl parathion could not meet the FQPA safety standards. The acute dietary risk to children ages one to six exceeded the acute population-adjusted dose (the amount that can be consumed safely in one day or less) by 880 percent. To lessen the high dietary risk to children, the EPA accepted voluntary cancellation of use on those crops that contribute most to children's diets. These canceled uses represent 90 percent of the dietary risk to children. Removing these crop uses brings the estimated dietary risk down to 78 percent of the reference dose, making the risk from food acceptable for children and all others in the United States.[45]

Restricted and Canceled Uses

The EPA registers pesticides and their use on specific pests and under specific circumstances. For example, "Pesticide A," registered for use on apples, may not be used legally on grapes; an insecticide registered for outdoor use may not legally be used inside a building. In some circumstances, use of a registered pesticide may be restricted to pesticide applicators with special training.

Over time, registered pesticides, or certain uses of registered pesticides, have been banned. These cancellations occur for various reasons: 1) voluntary cancellation by the registrant; 2) cancellation by the EPA because required fees were not paid; and 3) cancellation by the EPA because unacceptable risks existed that could not be reduced by other actions such as voluntary cancellation of selected uses or changes in the way the pesticide is used.[46]

Public Concern

The consequences of using pesticides for food production and the realization that some foods do contain pesticide residues are of paramount importance to today's health-conscious consumer. Specifically, the public continues to voice its concerns by ranking pesticide residue as one of the top five food safety issues. For example, public opinion polls indicate that in selecting produce, an important consideration is that foods are certified free of pesticide residues. The public's concern that consuming foods containing pesticide residues may adversely impact their health is critical.[47]

Why do consumers' attitudes toward pesticides and other health risks differ from those of health authorities?

According to the National Research Council, experts usually base their determinations of the seriousness of a risk on quantitative risk assessments or numerical probabilities. Consumers' risk perceptions tend to be based on qualitative attributes of risk as if the risk was previously known or unknown, voluntary or involuntary, or controllable or uncontrollable.[48]

Possible reasons for this attitude may be due in part to the uncertainty inherent to agrichemical use. For instance, it is impossible for any individual to quantify how much pesticide residue he or she is exposed to without explicit product labeling. Debates within the scientific community about the safety of insecticides and herbicides as well as specific events such as the Alar and Chilean grape incidents that have been widely publicized in the media have no doubt contributed to the concerns of consumers. The growing concern of residues in fresh produce could itself cause changes in consumer behavior in two ways: 1) increased demand for low-input agriculture with reduced pesticide residues, or 2) decreased demand for conventional fresh produce. For low-input agriculture to be marketed successfully, it will be necessary to determine whether consumer concern about pesticide residues has resulted in fundamental changes in consumer attitudes and behavior. An important foundation of this process is to assess which segments of the population are highly risk averse to pesticide usage.

While studies have found only modest variations in pesticide concern across different segments of the public, most have found that women are more likely than men to place pesticide residues as a top worry. Additionally, younger adults tend to show more concern over pesticide usage than older adults.[49] Respondents to a survey conducted by Cornell University felt that the lack of absolute evidence, the lack of simple precise documents addressing pesticide concerns, and conflicting information from experts all contributed to the complexity and level of public pesticide fears. Participants saw the pesticide dilemma as a long-term problem due to the vested interests of chemical manufacturers and the necessity of pest control, which conflict with public health and environmental fears.[50]

Safety information from the academic community was found to have the greatest likelihood of acceptance by consumers when compared to other information sources such as the government and the media. Other polls have indicated that 70 to 85 percent of the national population exhibits a medium to high degree of concern toward pesticide residue and pesticide usage. A study of four cities reported that 83 percent of respondents were risk averse to pesticide usage.[51] Another survey had 86 percent of respondents expressing concern for pesticide usage.[52] In another survey, more than 700 conventional and organic fresh produce buyers in the Boston area were asked about their perceived food safety risks. Responses indicated that consumers perceived relatively high risks associated with the consumption and production of conventionally grown produce compared to other public health hazards. For example, conventional and organic food buyers estimated the median annual fatality rate due to pesticide residues on conventionally grown food to be about fifty per million and 200 per million, respectively, which is similar in magnitude to the annual mortality risk from motor vehicle accidents in the United States. More than 90 percent of survey respondents also perceived a reduction in pesticide residue risk association with substituting organically grown produce for conventionally grown produce.[53] With sustainable and environmentally safer forms of agriculture likely to comprise a more significant share of the nation's food production, predicting which consumers are likely to have high concerns about synthetic pesticide residues should be beneficial to identifying those who are more likely to purchase low-input agriculture such as Integrated Pest Management (IPM) and organically grown produce.

The Organic Alternative: Reasons for Organic Food Growth

Expansion in organic food products requires a number of conditions, including sustained consumer interest and availability of organic-certified farmland on which to grow the crops in the United States. Consumer interest is paramount and food safety is likely to remain the most important issue for consumers. Food scares surrounding non-organically grown products (for example, BSE in cows or chemical and pesticide residues in fish) will drive consumers to seek out the perceived safety of organic alternatives. As long as organic foods are considered safe, more consumers will turn to them.

Three forces aid substantial organic growth: 1) rising consumer concern about the integrity of the food supply; 2) governmental standards that clearly define the meaning of the term "organic"; and 3) greater availability of organic foods through mainstream channels. It is significant that "organic" is a production claim, not a food safety or content claim. The term refers to the way in which food is grown and handled, and as a concept does not govern whether or not the food is safer or healthier than non-organic food. Nonetheless, consumers have embraced the term as signifying foods that are better for them or their families because foods grown organically are not grown with chemical fertilizers, pesticides, or hormones. Many consumers may be skeptical that organic foods are truly healthier or worth the extra expense or time spent finding them, but organic products now comprise 2 percent of total U.S. food sales. Organic food and beverages are gaining market share in mainstream channels, showing that both retailers and consumers are interested in food grown under organic conditions.

Claimed Benefits

In the food industry, defining the benefits of organic food is largely left to word of mouth, media coverage, and the promotional efforts of organic advocates. Major food and beverage corporations such as Kraft Foods, Heinz, Coca-Cola, Pepsi, Cargill, Unilever, General Mills, and Campbell's Soup have moved rapidly to acquire significant organic market share.[54] Still, the specific sales points of organics go largely unmentioned on product packaging and in mainstream media advertising. Claims of improved food quality are regularly used in conventional food marketing, with "low fat," "low sodium," "whole grain," "high fiber," "vitamin enriched," "no trans fat," and other commonly advertised benefits. By contrast, "certified organic" is generally left to stand on its own as self-explanatory, assisted only by general terms such as "natural." Meanwhile, consumer surveys have consistently identified food quality as the main reason for purchasing organic food. Higher nutritional value, no toxic residues from pesticides, and better taste are often cited, as is the positive impact of organic production on the environment.[55] Whether organic food actually delivers on these desires is controversial and the subject of scientifically inconclusive debate. The debate concentrates on a variety of specific and supposedly demonstrable characteristics that proponents have claimed make organic food production superior to conventional food production.

Risk for Children

Organic foods are good for children and they help preserve the rural environment. The U.S. Centers for Disease Control reports that one of the main sources of pesticide exposure for children comes from the food they eat. According to the FDA, half of produce currently tested in grocery stores contains measurable residues of pesticides. Laboratory tests of eight industry-leading baby foods revealed the presence of sixteen pesticides, including three carcinogens. According to "Guidelines for Carcinogen Risk Assessment," children receive 50 percent of their lifetime cancer risks in the first two years of life. In blood samples of children ages two to four, concentrations of pesticide residues were six times higher in children who ate conventionally farmed fruits and vegetables compared to those who consumed organic food.[56]

Organic Diets Versus Conventional Diets for Children

Two recent studies, one in 2003 and the other in 2005, both involving children in the Seattle, Washington, metropolitan area, confirm the value of organic diets vis-à-vis conventional menus for children. In the first study, preschool children ages two to four were the subjects. University of Washington researchers analyzed pesticide breakdown products (metabolites) in the children and found that those who ate organic fruits and vegetables had concentrations of pesticide metabolites six times lower than children who consumed conventional produce. The researchers compared breakdown concentrations of organophosphorus pesticides in the urine of thirty-nine urban and suburban children. Their findings point to a relatively simple way for parents to reduce their children's chemical loads—serve organic produce.

The authors focused on children's dietary pesticide exposure because children are at greater risk for two reasons: 1) they eat more food relative to body mass, and 2) they eat foods higher in pesticide residues, foods such as juices, fresh fruits, and vegetables. An earlier study had looked at pesticide metabolites in the urine of ninety-six urban and suburban children and found OP pesticides in the urine of all the children but one. The parents of the child with no pesticide metabolites reported buying exclusively organic produce.

The first Seattle study confirms what is already known about pesticide residues on conventional produce. Parents of young children have been warned to limit or avoid conventionally grown foods known to have high residues. The study's main conclusion—eating organic fruits and vegetables can significantly reduce children's pesticide loads—is information that parents can act on to reduce their children's risk. A secondary conclusion, that small children may be exceeding "safe" levels of pesticide exposure, is information that regulators should act on, at the very least, to reduce uses of these pesticides on food crops.[57]

In the second study, Dr. Chensheng Lu and his colleagues from Emory University, the University of Washington, and the Centers for Disease Control and Prevention (CDC) measured the exposure of two OPs, malathion and chorpyrifos, in twenty-three elementary students in the Seattle area by testing their urine over a fifteen-day period. The participants, ages three to eleven, were first monitored for three days on their normal diets. Then the researchers substituted most of the children's conventional diets with organic food items for five consecutive days. The children were then reintroduced to their normal foods and monitored for an additional seven days.

According to Dr. Lu, there was a "dramatic and immediate protective effect" against the pesticides until the conventional diets were reintroduced. While consuming organic food, most of the children's urine samples contained zero concentration for the malathion metabolite. However, once the children returned to their normal diets, the average metabolite concentration increased to 1.6 parts per billion with a concentration range from five to 263 parts per billion. A similar trend was observed for chlorpyrifos, as the average chlorpyrifos metabolite concentration increased from one part per billion during the organic diet days to six parts per billion when children resumed eating conventional food.[58]

Increased Antioxidant Levels

Other recent research has demonstrated that organic farming methods have the potential to elevate average antioxidant levels, especially in fresh produce. One study determined that, on average, antioxidant levels were about 30 percent higher in organic foods compared to conventional foods grown under the same conditions. This is particularly useful for people who wish to consume higher levels of antioxidants in fresh fruits and vegetables without additional caloric intake. The USDA recommends higher daily consumption of fruits and vegetables, especially those that are antioxidant rich. Consumers who seek out foods high in antioxidant content can meet recommended antioxidant intake levels with less than 10 percent of their daily caloric intake. The report reviews, among other data, fifteen quantitative comparisons of antioxidant levels in organic versus conventional fruits and vegetables. Organically grown produce had higher levels in thirteen out of fifteen cases. On average, the organic crops contained about one-third higher antioxidant and/or phenolic content than comparable conventional produce.

Several studies found levels of specific vitamins, flavonoids, or antioxidants in organic foods to be two to three times the level found in matching samples of conventional foods. In studies making direct comparisons of levels of antioxidants in organic versus conventional produce, higher levels are often found in organic produce but the converse is rarely true.

There is evidence that several core practices on organic fruit and vegetable farms, such as use of compost, cover crops, and slow release forms of nitrogen, can increase antioxidant and polyphenol content compared to conventional practices that depend on commercial fertilizers and pesticides. The prohibition of pesticides in organic farming provides additional benefits to consumers who choose organic.

Furthermore, there are significant differences between some of the food processing methods and technologies used in manufacturing conventional foods in contrast to those allowed and used in producing organic processed foods. Some of these differences are known to have an impact on antioxidant levels. For example, the synthetic chemical hexane is often used in extraction of oil from crops in conventional oil processing plants, but is prohibited in organic oil processing. Hexane is known to promote the removal of certain antioxidants.

High-temperature and high-pressure processing technologies also tend to remove significant portions of the antioxidants present in fresh foods. Organic processing plants often use lower-pressure, cold-pressing methods to extract juices and oils. The resulting food products are generally richer in flavor and retain more nutrients, including antioxidants.

Though there is much to learn, the current state of science supports the conclusion that organic farming methods can and often do result in higher antioxidant levels in fruits and vegetables. This health benefit for consumers joins the list of other well-documented reasons to buy organic, including the reduction of farmworker and consumer exposures to pesticides, the impacts of pesticides on the environment, and the prevention of problems associated with hormone and antibiotic use in livestock farming. Many consumers report that they enjoy the richer flavors in organic food and instinctively sense that organic foods are better for them; research confirms that there are good reasons to focus additional scientific resources on gaining a more comprehensive understanding of the taste and health benefits associated with elevating average antioxidant levels in food.[59]

Fewer Pesticide Residues

Extensive and highly sensitive pesticide testing carried out by the USDA shows that conventional fresh fruits and vegetables are: 1) three to more than four times more likely on average to contain residues than organic produce; 2) eight to eleven times more likely to contain multiple residues than organic samples; 3) shown to contain residues at levels three to ten times higher, on average, than corresponding residues in organic samples.

For many people, consumption of organic fruits and vegetables on most days will virtually eliminate dietary exposure to pesticides and, in turn, reduce the frequency and magnitude of one risk factor that can contribute to a variety of diseases and health problems. The opportunity to nearly eliminate pesticide exposure via diet by consuming organic food is borne out by extensive testing in the United States and other countries. The pesticide risk reduction benefits of consuming certified organic apples, pears, peaches, strawberries, cherries, celery, spinach, and sweet bell peppers are particularly significant, especially for women of childbearing age, infants, and children.[60]

Pesticide Residues in Organic Food

Certified organic food may not be treated with synthetic pesticides, so why do residues of synthetic pesticides sometimes appear on organic food? Pesticides are ubiquitous and mobile across agricultural landscapes. Most organic samples contain low levels of pesticides that were sprayed on nearby conventional crops. Pesticides applied on conventional crop acreage sometimes drift in the air and settle onto the plants growing on nearby organic farms. When some pesticides are applied using airplanes, as little as 25 percent of the applied pesticides settle on the target crops, while three-quarters drifts off site. When pesticides are applied using ground equipment on days with modest to moderate winds, losses of 25 percent or more via drift are common. Pesticides also sometimes travel in fog. Irrigation water also moves across agricultural landscapes, picking up pesticide contamination along the way.

When residues of synthetic pesticides do show up on organic foods, the levels are on average lower than corresponding residues in conventional food. The National Organic Program (NOP) rule calls upon certifiers to investigate cases in which a residue of a synthetic pesticide appears on organic food greater than 5 percent of the applicable European Partners for the Environment (EPE) tolerance. NOP adopted this policy to prevent organic farmers from losing certification over incidental environmental contamination with pesticides not actually applied on their crops.[61]

Organic Certification

In 2002, when the USDA adopted the National Organic Standard that spells out what farmers and food processors must (and must not) do to be certified "organic," the organic industry already had a long history of relying on third-party certifiers to ensure the integrity of their products and practices. Under this system, a state-run or accredited private agency (the third party) evaluates farmers and processors to see if they conform to the standards of the National Organic Program. Those who can then market their products as "USDA Certified Organic" and display the official USDA organic seals on their packaging.

In essence, certification is largely about integrity—assuring that the buyer is getting what he or she is paying for. Thus, certified organic production means production by approved organic methods, with additional pains taken to eliminate contamination by prohibited materials and commingling with conventional products. There is a common misconception that certified organic means "pesticide-residue-free." Consumers have a right to expect little or no pesticide residue on certified organic crops because none are used in their production. However, ours is a dirty world in which pesticides and their breakdown products are omnipresent. This is only to be expected in a national farm system where more than 99 percent of all applied farm chemicals miss the target organism.[62]

Organic Labeling

National organic standards address the methods, practices, and substances used in producing and handling crops, livestock, and processed agricultural products. Although specific practices and materials used by organic operations may vary, the standards require every aspect of organic production and handling to comply with the provisions of the Organic Foods Production Act.

Labeling requirements under the national standards apply to raw foods, fresh products, and processed foods that contain organic ingredients and are based on the percentage of organic ingredients in a product. Agricultural products labeled "100 percent organic" must contain (excluding water and salt) only organically produced ingredients. Products labeled "organic" must consist of at least 95 percent organically produced ingredients. Products labeled "made with organic ingredients" must contain at least 70 percent organic ingredients. Products with less than 70 percent organic ingredients cannot use the term organic anywhere on the principal display panel, but they may identify the specific ingredients that are organically produced on the ingredients statement on the information panel. In a processed product labeled as "organic," all agricultural ingredients must be organically produced unless the ingredients are not commercially available in organic form. The USDA organic seal—the words "USDA organic" inside a circle—may be used on agricultural products that are "100 percent organic" or "organic." A civil penalty of up to $10,000 can be levied on any person who knowingly sells or labels as organic a product that is not produced and handled in accordance with these regulations.[63]

The Booming Market for Organics

The number of Americans who tried organic foods jumped to 65 percent in 2005, compared to 54 percent in 2003 and 2004. One quarter of 1,000 people responding to survey said they were consuming more organics than they had the year before. In the annual survey, 10 percent of respondents said they consume organic foods several times per week, up from just 7 percent in 2004.

Americans are buying organic foods and beverages for a variety of reasons. The top three are: avoidance of pesticides (70.3 percent), freshness (68.3 percent), and health and nutrition (67.1 percent). More than half (55 percent) buy organic to avoid genetically modified foods. Also, more than half of all respondents agreed that organic foods and beverages are "better for my health" (52.8 percent) and better for the environment (52.4 percent).

The survey unveiled significantly higher taste and quality ratings from Americans who regularly consume organic foods and beverages. Fresh fruits and vegetables remain overwhelmingly the most frequently purchased category of organic foods at 73 percent. Produce is followed by non-dairy beverages (32 percent), bread or baked goods (32 percent), dairy items (24.6 percent), packaged goods such as soup or pasta (22.2 percent), meat (22.2 percent), snack foods (22.1 percent), frozen foods (16.6 percent), prepared and ready-to-eat meals (12.2 percent), and baby food (3.2 percent). One quarter of respondents said they purchase organic foods at natural foods supermarkets, while 18 percent shop for organics at farmers' markets.

The main barrier to purchasing continues to be price; almost three-quarters (74.6 percent) of respondents said the cost of organic food and beverages is the main reason they do not consume more. Other reasons Americans are not consuming more organics, according to the survey, include availability (46.1 percent) and loyalty to non-organic brands (36.7 percent).[64]

Despite the lack of national organic standards before 2002, sales of organic products have increased on average by 20 percent annually since 1990. Medium-term growth forecasts for U.S. organic markets is in the range of 20 to 30 percent.[65] Experts predict that the organic industry's share of the U.S. food market is expected to grow from about 2 percent to roughly 3.3 percent by the end of the decade. Organic food sales in the United States are projected to reach $30.7 billion in 2007, driven largely by double-digit growth in the meat and meat products industry. Sales of organic meat and meat products are expected to grow from $547 million in 2002 to $3.86 billion in 2007.[66]

Demand Outstrips Supply

Growth in the U.S. organic market is being stunted by undersupply, resulting in shelves remaining empty, companies withdrawing from the market, and others looking internationally to supplement supply needs. The sectors hardest hit by supply shortages are the organic orange juice, meat, and dairy sectors. The low number of organic livestock producers in the United States has also resulted in the organic meat industry experiencing undersupply for a number of years, with American producers resorting to imported organic beef from Australia and Latin American countries.

Increasing quantities of organic fruits, vegetables, grains, seeds, beans, and herbs are being imported into the United States. Finished products are also being imported to meet consumer demand for all things organic. As consumers increasingly opt for healthier products, the organic industry is set to enjoy solid growth. According to a recent study, by 2025 organic products will be considered commonplace. Yet, a shortage of supply is stifling sales. Nearly all market sectors would grow at much higher rates if sufficient supplies were available.

Once a net exporter of organic products, the United States now spends more than $1 billion a year to import organic foods, according to the USDA; the ratio of imported to exported products is now about eight to one. This situation is occurring despite the fact that about 10,000 American farmers have made the transition to organic food production on about 2.3 million acres of land. Hopefully, the number of farmers converting from conventional to organic farming will grow more rapidly, thereby easing the supply shortages.[67]

Corporate Inroads

Today a significant, and growing, percentage of organic foods are produced by large corporations more often associated with the predations of agribusiness than with the ideals of sustainable farming. The increasing presence of conventional food processors in the organic industry is raising debate among farmers, shoppers, and consumer advocates about whether the values of organic agriculture and the motives of big business can coexist.

Figures supplied by the Organic Consumers Association reveal the degree to which conventional food processors have penetrated the organic market. General Mills owns the organic brands Cascadian Farms and Muir Glenn. Heinz holds a 20 percent equity share in food distributor Hain, which owns Rice Dream soy milk, Garden of Eatin', Earth's Best, and Health Valley, along with fifteen other organic brands. Kellogg owns Sunrise Organic, while Kraft owns Boca Foods, maker of the popular vegetarian Boca burgers. The largest organic seed company, Seeds of Change, is controlled by M&M/Mars. Small, local organic operations simply do not have the reach to coordinate nationwide distribution. Bigger companies do have the expertise in getting products to shelves across the country. As the market for organics has grown, they have stepped in to fill that role.

Farmers, advocates, and ordinary shoppers share the view that the mainstreaming of organics carries both benefits and risks. On one hand, more organic foods are available to people than at any time since the advent of the industrial food age, and this should have very real benefits for public health and the environment. On the other hand, some fear that corporate giants don't really believe in the values of sustainable farming and that, in the long run, their participation in the industry will dilute the very meaning of the term "organic."[68]

Organic Standards Are Endangered

Organic standards, which ban synthetic fertilizers, antibiotics, hormones, pesticides, genetically engineered ingredients, and irradiation, are good for farming, the environment, and public health. The organic seal is vitally important in stores, where the consumer is several steps removed from the farmer. "Organic" is a legal guarantee that food meets certain standards.

The Organic Trade Association, a food industry group whose members include national organic brands such as Kraft, Dean Foods, and General Mills, is seeking to dilute organic standards. If Big Organic gets its way, xanthan gum (an artificial thickener), ammonium bicarbonate (a synthetic leavening agent), and ethylene (a chemical that ripens tomatoes and other fruit) will be permitted in products labeled organic, despite a 2005 court ruling indicating they are not acceptable. Whatever the outcome of legal maneuvering, consumers should look beyond the organic label and seek out producers who exceed the federal rules. If the organic label loses its meaning, farmers with higher standards will have to devise new ones. The next generation of labels could read "grass-fed" butter and "pastured" pork. These foods, and others raised with ecological and humane methods, are superior to industrial organic foods. Agriculture departments may never tell you that, but smart farmers will.[69]

Failure of a Legislative Challenge

Interestingly, U.S. national organic standards were put to the test only several months after implementation. The good news: the organic sector was able to uphold the integrity of the standards.

The challenge came in February 2003, in the form of one long sentence, Section 771, hidden in the Omnibus Appropriations Bill. The rider, although not overturning the national organic standards, would have undermined organic standards by failing to fund USDA's enforcement of the requirement of 100 percent organic feed for all livestock. In effect, if left to stand, it would have opened the door to lesser requirements for livestock feed, and made it impossible for consumers to trust the organic label on organic livestock-derived products, from meat and eggs to dairy products.

This raised the ire of those already willing to meet the 100 percent requirement, as well as the Organic Trade Association, affiliated organizations, organic food companies, and consumers. Some legislators, who had been on the ground floor when the

Organic Foods Production Act of 1990 was enacted, fought to overturn this rider. They were joined by others who might not have actually supported organic agriculture, but believed it was important to let the new regulation stand and be enforced.

In addition, Secretary of Agriculture Ann M. Veneman also stepped up to defend the national organic standards. In the final analysis, the USDA was willing to back the National Organic Program. The bottom line: consumers need to be able to trust a label, and the new rule needed to be given a chance to work. There is another win that can be traced in part from this incident: growing congressional awareness of the importance of organic agriculture and products. As a result, the U.S. House of Representatives has established a formal Organic Caucus, and the U.S. Senate has in place an informal organic working group. These developments signal a "coming of age" for the organic sector in the legislative arena.[70]

A Final Caveat

Capitalism is based on incentives and demand, demand created for the products and the process and benefits of organic farming, sustainability of our resources, and the health of our bodies. This system was not created in a vacuum, however, nor does it exist without structure. That structure takes the form of massive government subsidies, a model that is duplicated across Europe in countries that care about their local small farmers. Unfortunately, subsidies in the United States typically don't go to the local small farm, but instead to large regional corporations that impose non-sustainable practices and poisons on smaller farms and the environment.

Chemical manufacturers and some farm organizations feel that dietary dangers from pesticides have been exaggerated, while some consumer groups and scientists believe that the danger has been understated. We do not know for sure who is right or wrong on this issue. In the meantime, we prefer to be on the side of safety.

Notes

1. W. J. Kroll, T. L. Arsenault, H. M. Pylypiw Jr., and M. J. L. Mattina, "Reduction of Pesticide Residues on Produce by Rinsing," *Journal of Agriculture and Food Chemistry* 48 (2000): 4,666–4,670.

2. Agricultural Marketing Service, *Pesticide Data Program: Annual Summary (Calendar Year 2000)* (Washington, D.C.: U.S. Department of Agriculture, 2000).

3. Agricultural Marketing Service, *Pesticide Data Program: Annual Summary (Calendar Years 1993–1998)* (Washington, D.C.: U.S. Department of Agriculture, 1999).

4. Office of Pesticide Programs, *Setting Tolerances for Pesticide Residues in Foods* (Washington, D.C.: Environmental Protection, Office of Pesticide Programs, 2000).

5. Pesticide Monitoring Program, *Residue Monitoring (Calendar Years 1993–1999)* (Washington, D.C.: U.S. Food and Drug Administration, Pesticide Monitoring Program, 2000).

6. Ibid.

7. Brian P. Baker, Charles M. Benbrook, Edward Groth III, and Karen Lutz Benbrook, "Pesticide Residues in Conventional, IPM-Grown, and Organic Foods: Insights from Three U.S. Data Sets," *Food Additives and Contaminants* 19 (5) (May 2002): 427–446.

8. R. Wiles and C. Campbell, *Pesticides in Children's Food* (Washington, D.C.: Environmental Working Group, 1993).

9. National Research Council, *Pesticides in the Diets of Infants and Children* (Washington, D.C.: National Academies Press, 1993).

10. *Handbook of Pediatric Environmental Health* (Elk Grove, IL: American Academy of Pediatrics, 1999): 203.

11. National Research Council, *Pesticides in the Diets.* Op. cit.: 305.

12. National Research Council, *Regulating Pesticides in Foods: The Delaney Paradox* (Washington, D.C.: National Academy Press, 1987).

13. *Federal Register 64* (August 6, 1999): 42,945–42,947.

14. Linda Jo Schierow, *Pesticide Residue Regulation: Analysis of Food Quality Protection Act Implementation* (Washington, D.C.: Congressional Research Service, November 4, 2002).

15. Adam Kirshner, "New U.S. Food Quality Protection Act: Does it Protect Consumers?" *Global Pesticide Campaigner* 6 (3) (September 1996).

16. *Setting Tolerances for Pesticide Residues in Foods* (Washington, D.C.: Environmental Protection Agency, June 13, 2003).

17. U.S. Department of Health and Human Services, *Pesticide Residue Monitoring Database Users' Manual* (Washington, D.C.: U.S. Food and Drug Administration, Center for Food Safety and Applied Nutrition, April 2003).

18. Brian P. Baker et al. Op. cit.

19. Ibid.

20. *Pesticides—Summary of FQPA Amendments to FIFRA and FFDCA* (Washington, D.C.: U.S. Environmental Protection Agency, February 7, 2005).

21. Phillip Paarlberg and Phillip Abbott, *U.S. on Verge of Becoming Net Agricultural Importer* (West Lafayette, IN: Purdue University, September 19, 2003).

22. *Toxicity Testing Fact Sheet* (Washington, D.C.: National Institute of Environmental Health Sciences, January 2004).

23. *Putting Children First: Making Pesticide Levels in Food Safer for Infants and Children* (Washington, D.C.: Natural Resources Defense Council, April 1998).

24. Consumer Health Staff, "Pesticides on Fruits and Vegetables," *Consumer Health* 22 (7) (July 1999).

25. Fact Sheet #25, *Pesticide Residue Monitoring and Food Safety* (Ithaca, NY: Cornell University, March 1999).

26. Center for Food Safety and Applied Nutrition, *Pesticide Program: Residue Monitoring, 1996* (Washington, D.C.: U.S. Food and Drug Administration, 1998).

27. *Nowhere to Hide: Persistent Toxic Chemicals in the U.S. Food Supply* (San Francisco, CA: Pesticide Action Network North America, 2000).

28. *Overexposed: Organophosphate Insecticides in Children's Food* (Washington, D.C.: Environmental Working Group, 1998): 1–3.

29. Office of Pesticide Programs, *Assigning Values to Non-Detected, Non-Quantified Pesticide Residues in Human Health Food Exposure Assessments* (Washington, D.C.: Environmental Protection Agency, March 2000).

30. "The Process of Tolerance Reassessment," *Agronomy News* (May 1999).

31. National Research Council, *Regulatory Pesticides*. Op. cit.

32. A. R. Bates and S. Gorbach, "Recommended Approach to the Appraisal of Risks for Consumers from Pesticide Residues in Crops and Food," *Pure and Applied Chemistry* 59 (1987): 611–624.

33. America's Children and the Environment, *Pesticides on Foods Frequently Consumed by Children* (Washington, D.C.: Environmental Protection Agency, n.d.); D. Wallinga, *Putting Children First: Making Pesticide Levels in Food Safer for Infants and Children* (New York, NY: Natural Resources Defense Council, 1998): 15.

34. *Overexposed*. Op. cit.

35. Consumers Union Report, press release, *Pesticide Residues Still Too High in Children's Foods* (Washington, D.C.: Consumers Union, June 6, 2000).

36. *Case Study: Organochlorine Pesticides* (San Francisco, CA: Pesticide Action Network North America, n.d.).

37. Consumers Union et al. Op. cit.

38. Consumers Union et al. Op. cit.

39. Center for Food Safety and Applied Nutrition, *Channels of Trade Policy for Commodities with Residues of Pesticide Chemical, for Which Tolerances Have Been Revoked, Suspended, or Modified by the Environmental Protection Act Pursuant to Dietary Risk Considerations* (Washington, D.C.: U.S. Food and Drug Administration, n.d.).

40. Linda Jo Schierow. Op. cit.

41. Helpguide.org, "Organic Foods: Guide to Pesticides, GMOs, Food Irradiation, and Eating Well On A Budget" (2006): 1.

42. Jane Kay, "Pesticide Threat to Babies Linked to Enzyme Levels: Researchers Find Them Much More at Risk than Adults," *San Francisco Chronicle*, March 3, 2006.

43. "Residue Monitoring," *Journal of Association of Official Analytical Chemists* 75 (1) (1991): 135A–157A.

44. J. E. Delmore and A. D. Appelhans, "Detection of Agricultural Chemical Residues on Plant Leaf Surfaces with Secondary Ion Mass Spectrometry," *Biological Mass Spectrometry* 20 (5) (1991): 237–246.

45. *EPA Bans Pesticide Methyl Parathion: Fact Sheet* (Washington, D.C.: Environmental Protection Agency, August 4, 1999).

46. *Regulating Pesticides* (Washington, D.C.: Environmental Protection Agency, February 1, 2006).

47. Fred Whitford, Linda Mason, and Carl Winter, *Profile of Concern* (West Lafayettte, IN: Purdue University Pesticide Program, n.d.).

48. National Research Council, *Improving Risk Communication* (Washington, D.C.: National Academy Press, 1989).

49. Riley E. Dunlap and Curtis E. Beus, "Understanding Public Concerns About Pesticides: An Empirical Examination," *The Journal of Consumer Affairs* 26 (2) (1992): 418–438.

50. Nancy Ostiguy et al., "Improving Communication About Risks Associated with Residues of Agrichemicals in Produce," *Agricultural Economic Bulletin 90–23* (Ithaca, NY: Cornell University, 1990).

51. J. A. Zellner and B. L. Degner, "Consumer Willingness to Pay for Food Safety," paper presented at the Southern Agricultural Meeting, Nashville, TN, 1989.

52. T. Zind, "Fresh Trends 1990: A Profile of Fresh Produce Consumers," *The Packer-Focus 1989–1990* (Overland Park, KS: Vance Publishing Co., 1990).

53. P. R. D. Williams and J. K. Hammitt, "Perceived Risks of Conventional and Organic Produce: Pesticides, Pathogens, and Natural Toxins," *Risk Analysis* 21 (2) (April 2001): 319–330.

54. Phil Howard, *Corporate Industry Structure: 2005* (Santa Cruz, CA: University of California, Center for Agroecology and Sustainable Food Systems, 2005).

55. Carol Raab and Deana Grobe, "Consumer Knowledge and Perceptions About Organic Food," *Journal of Extension* (August 2005).

56. EPA Guidelines for Carcinogen Risk Assessment, *Federal Register* 51 (185) (September 24, 1986): 33,992–34,003.

57. Cynthia L. Curl, Richard A. Fenske, and Kai Elgethun, "Organophosphorus Pesticide Exposure of Urban and Suburban Pre-School Children with Organic and Conventional Diets," *Environmental Health Perspective* 3 (3) (March 13, 2003): 337–392.

58. Chensheng Lu et al., "Organic Diets Significantly Lower Children's Dietary Exposure to Organophosphorus Pesticides," *Environmental Health Perspectives* 114 (2006): 260–263.

59. *Certain Organic Farming and Food Processing Techniques Can Increase Antioxidant Levels* (Finland, MN: Organic Consumers Association, January 26, 2005).

60. D. Bourn and J. Prescott, "A Comparison of the Nutritional Value, Sensory Qualities, and Food Supply of Organically and Conventionally Grown and Produced Foods," *Critical Reviews in Food Science and Nutrition* 42 (1) (2002): 1–34.

61. D. W. Lotter, R. Seidel, and W. Liebhardt, "The Performance of Organic and Conventional Cropping Systems in an Extreme Climate Year," *American Journal of Alternative Agriculture* 18 (2003): 146–154.

62. D. Pimental and L. Levitan, "Pesticides: Amounts Applied and Amounts Reaching Pests," *BioScience* 36 (1986): 86–91.

63. Leslie A. Duram, "Factors in Organic Farmers' Decision Making: Diversity, Challenge, and Obstacles," *American Journal of Alternative Agriculture* 14 (1) (1991): 2–10.

64. "Organics Blooming: 65 Percent of Americans Tried Organic Foods in 2005," *The Progressive Grocer* (November 21, 2005).

65. C. R. Greene, "U.S. Organic Emerges in the 1990s: Adoptions of Certified Systems," *Agricultural Information Bulletin No. 770* (Washington, D.C.: U.S. Department of Agriculture, Economic Research Series, 2001).

66. Barbara Haumann, "United States 7.6.1" in Helga Willer and Minou Yussefi, eds., *The World of Organic Agriculture—Statistics and Emerging Trends—2004* (Bonn, Germany: International Federation of Organic Agriculture Movements, 2004): 155.

67. "Organic Market Stunted by Undersupply," *Organic Monitor* (Finland, MN: Organic Consumers Association, December 19, 2005).

68. Jason Mark, "Conventional Food Processors in the Organic Industry Raise Debate about the Value of Organic Agriculture and the Motives of Big Business," *The Monthly* (October 1, 2004).

69. Nina Planck, "Beyond USDA Organic: Buying Local, Organic, and Grass-Fed," *The New York Times*. November 23, 2005.

70. Barbara Haumann. Op. cit.: 153–154.

Four

Pesticides in Schools

Unfortunately, they don't envision a helicopter flying a half a mile away putting out pesticides. They don't envision that their child's school is gonna be right across the street from a lemon orchard.

—Susan Johnson[1]

Introduction

One of the most important ways to protect our children's health is to prevent their exposure to pesticides and other toxins that may be used in their schools. Children are far more sensitive to low concentrations of toxic chemicals because of their developing organs and high metabolism. Pesticides are often used for ridding school buildings and areas of rodents, insects, and other pests. But they only work temporarily and must be reapplied. However, the poisons found in pesticides may harm more than just pests. Children already have proportionally more pesticide exposures than adults. Pesticides have been linked to certain cancers, damage to the central nervous system, and neurological and behavior problems, as well as acute poisoning. No pesticides have been tested specifically for threats to children or in combination with other chemicals.

Schools should be environmentally healthy places for children to learn, for teachers to teach, and for other school employees to work. Our society suffers when schools become so run down and toxic that they become a stress to the body's systems rather than an inspiration to young minds.

Likewise, playgrounds should be places where children can play without risk of being exposed to pesticides, contaminated play structures, or other health hazards. Parents should not have to worry about much more than sprained ankles and scuffed kneecaps when they let their children play in such areas.

The simple fact is that while parents can exert control over the chemicals they use in their homes, the same doesn't necessarily hold true for the schools and playgrounds where their children spend much of their time. Every day, school and playground environments expose children to high levels of toxic substances, from industrial-strength pesticides to harmful building and cleaning materials to playground equipment made of arsenic-treated wood.

Schools and their kitchens, cafeterias, athletic fields, playgrounds, classrooms, and offices are regularly treated with a variety of pesticides. An increasing body of scientific data on the potentially harmful effects of pesticide exposure on people and the environment rightfully raises concerns about the broad use of these toxic substances. Children spend 30 to 50 percent of their waking hours in school, nine months of the year, making a healthy school environment all the more vital to their growth and development.

The GAO reported in 2000 that it could find no credible evidence on how much pesticide is used in the nation's 110,000 public schools, how often students are exposed to dangerous chemicals at school, or what the health effects are. Pesticide opponents estimate there are some fifty insecticides, herbicides, and fungicides commonly used in and around schools. Many herbicides applied to school grounds may leave persistent residues in soil for weeks, months, or even years.[2]

Lack of Knowledge

Most parents know that their children are exposed to pesticides on foods. But many may not be aware of the quantity or pervasiveness of other non-food pesticide exposures their children encounter in a typical day, or they may not understand the risk of these cumulative exposures. Parents may not be aware of pesticides in schools.

Many people assume that schools are environmentally safe places for children to learn. It often takes a pesticide poisoning, repeated illnesses, or a strong advocate to alert a school district to the acute and chronic adverse health effects of pesticides and the viability of safer pest management strategies. Schools that have chosen to adopt such strategies, such as an Integrated Pest Management (IPM) program, use alternatives to the prevailing chemical-intensive practices because of the health risks such practices pose to children and other school users.

What Organizations Say about IPM

The American Public Health Association, the National Association of School Nurses, and the National Parent-Teachers Association support effective alternative pest control methods such as IPM in these comments:

"In managing pests, the emphasis should be placed on minimizing the use of broad spectrum chemicals, and on maximizing the use of sanitation, biological controls, and selective methods of application."

"A healthy school environment is essential. All students and staff have a right to learn and work in a healthy school environment, safe from air pollution, radiation, sound and mechanical stress, and chemical exposures."

"National PTA supports efforts [at IPM implementation] at the federal, state, and local levels to eliminate the environmental health hazards caused by pesticide use in and around schools."[3]

The Nature of Exposure

Despite the widespread use of pesticides, often only limited precautions are taken in terms of warning signs. While pesticides may be used in all parts of a school, most often in kitchens, cafeterias, and on the grounds outdoors, staff, students, and parents often have no way of knowing when they may be exposed.

Depending upon the pesticide, the target pest, the site to be treated, and other factors, pesticides may be applied as powders, pellets, liquid sprays, fogs, or mists, or mixed with some sort of bait to attract the pest. In each case, human exposure may occur. It may be at the application site or at other locations. The pesticide may be carried by airborne drift, surface runoff, or tracking, as well as by routine mopping or sweeping. Pesticides do not disappear immediately after application. They may take days, weeks, or even months to break down outdoors. Indoors, away from sunlight and soil bacteria that often help in their breakdown, pesticides may persist far longer. Even natural degradation is not always the answer; for some pesticides, the natural breakdown product is even more toxic than the original pesticide.

Also, pesticides are often used in areas of a school where exposure risks are heightened. For example, pesticides used in the cafeteria could end up in the food students eat. Pesticides on a gymnasium floor could end up on the hands and ultimately in the body via the mouth, the eyes, or other entry points. Outdoors, use of pesticides on grass and plants in and around a school can unnecessarily expose children to toxins.

Chemical sensitivity may be a reaction to pesticides. Even though pesticides are applied carefully, they can travel on air currents to affect chemically sensitive people. Liquid pesticides are volatile and have been shown to move from the application site to areas where no pesticides have been applied.

Surface treatments from a hand-held, compressed-air pump sprayer increase the risk of exposure to airborne pesticides. This exposure may trigger reactions that could be life threatening to sensitive individuals. Schools have the responsibility to provide safe environments without the risk of exposure to pesticides. A school district may also be exposing itself to legal liability if a child's health problems are traced to exposure in and around a school.

Each school must decide if pesticides will be stored on campus. If schools store pesticides on campus, control measures must be strictly followed to limit and document their access and use in order to reduce any risk of accidental poisoning. Many schools do not store pesticides properly. It is common to find improperly stored pesticides that are accessible to children in the classroom. Pesticides have been found in sink-based cabinets, on shelves, and even on teachers' desks. The improper storage of pesticides is an accident waiting to happen. One of the primary responsibilities of a school is to decide on the proper storage and handling procedures for on-site pesticides. Again, legal liability lingers in the background for a school district if a child is poisoned due to improper pesticide storage.[4]

A 1999 survey of Connecticut schools found that 87 percent of the state's school districts that responded (seventy-seven of 147 districts) sprayed pesticides inside school buildings; 32 percent sprayed pesticides routinely regardless of whether there was a pest problem.[5] A 1998 survey of California school districts revealed that 93 percent of forty-six districts responding used pesticides.[6] An earlier survey taken in 1991 of 261 New York schools, indicated that 87 percent used pesticides.[7] Some commonly used insecticides, such as pyrethroids, that stimulate nerves, causing hyperexcitability. They are also associated with asthma. Certain insecticides, herbicides, and fungicides are linked to cancer. The commonly used weed killer 2,4-D has been linked to non-Hodgkin's lymphoma in scientific studies.[8]

Three Pesticides to Avoid

Among the most dangerous toxins is Roundup (glyphosate), which kills all green plants that it touches (users are advised to avoid treatment areas for twenty-four hours). Another is diazinon, used for killing insects in lawns (it has a warning to keep away from edible plants because of its high degree of toxicity). Some schools have opted to pull weeds by hand, thus eliminating the need for spraying.[9] Another pesticide often applied, Dursban (chlorpyrifos), has a half-life greater than thirty days. As a result, a classroom carpet can become a continuing reservoir of pesticide exposure, long after the application appears to have dried. The pesticide vapors build up into an invisible, odorless, toxic chemical soup that is capable of causing a cascade of toxic reactions for years to come. The California Department of Health Services estimated the amount of chlorpyrifos to which a child would be exposed one day after an indoor application. The estimate was based on the amount that the child would breathe added to the amount the child would absorb through the skin. The estimate was *more than 1,700 times the acceptable daily intake* established by the World Health Organization. The aerial drift of one droplet of pesticide on a calm day was reported at twenty-four miles. On a windy day, aerial drift has been traced halfway around the globe. The EPA reports that "chlorpyrifos has ... been associated with chronic effects in humans, including chronic neurobehavioral effects and multiple chemical sensitivity. Neurobehavioral effects reported include persistent headaches, blurred vision, unusual fatigue or muscle weakness, and problems with mental function including memory, concentration, depression, and irritability."[10] Studies suggest that human health effects may occur even in the absence of measurable depression of the enzyme cholinesterase, which the EPA usually considers the most sensitive measure of exposure to organophosphates such as Dursban.[11]

An exhaustive investigation by the Greater Boston Physicians for Social Responsibility (GBPSR) examined the contribution of toxic chemicals to neurodevelopmental, learning, and behavioral disabilities in children. The Boston physicians' report indicated that these disabilities are clearly the result of complex interactions among genetic, environmental, and social factors that impact children during vulnerable periods of development. Toxic exposures deserve special scrutiny because they are preventable causes of harm.

An epidemic of developmental, learning, and behavioral disabilities among children has become evident. It is estimated that nearly twelve million children (17 percent) in the United States under age eighteen suffer from one or more learning, developmental, or behavioral disabilities. According to conservative estimates, attention deficit hyperactivity disorder (ADHD) affects 3 to 6 percent of all schoolchildren, though recent evidence suggests the prevalence may be as high as 17 percent. The number of children taking the drug Ritalin for this disorder has roughly doubled every four to seven years since 1971 to reach an estimated 1.5 million. Learning disabilities alone may affect approximately 5 to 10 percent of children in public schools. The number of children classified with learning disabilities in special education programs increased 191 percent from 1977 to 1994.

Pesticides

Animal tests of pesticides belonging to the commonly used organophosphate class of chemicals show that small single doses on a critical day of development can cause hyperactivity and permanent changes in neurotransmitter receptor levels in the brain. Chlorpyrifos, one of the most commonly used organophosphates, decreases DNA synthesis in the developing brain, resulting in deficits in cell numbers. Some pyrethroids, another commonly used class of pesticides, also cause permanent hyperactivity in animals exposed to small doses on a single critical day of development. Children exposed to a variety of pesticides in an agricultural community in Mexico show impaired stamina, coordination, memory, and capacity to represent familiar subjects in drawings.

These trends may reflect true increases, improved detection, better reporting, improved record keeping, or some combination of these factors. Whether new or newly recognized, these statistics suggest a problem of epidemic proportion.[12]

The JAMA Study

A recent widely reported study in the *Journal of the American Medical Association* underscored the risks of pesticide use in and around the nation's schools. Analyzing 2,593 reported pesticide poisonings in schools and childcare centers between 1998 and 2002, the study reported several troubling findings: incidence rates among children increased significantly from 1998 to 2002; drifting pesticides applied off site were responsible for 31 percent of reported poisonings; and insecticides and disinfectants were the pesticides most frequently at fault. The study's authors noted that no federal requirement limits pesticide exposures at childcare centers or elementary or secondary schools. They stressed their results should be considered low estimates of the magnitude of the problem because many cases of pesticide poisoning are likely not reported to surveillance systems and poison control centers.

The study examined state surveillance data from the National Institute for Occupational Safety and Health's Sentinel Event Notification System for Occupational Risks pesticides program, from the California Department of Pesticide Regulation, and from a national database of calls made to poison control centers compiled by the American Association of Poison Control Centers' Toxic Exposure Surveillance System. The study reported incidence rates of pesticide-related illness of 7.4 cases per million for students and 27.3 cases per million for employees. It also emphasized that individuals had to seek medical care and report exposures in order to be counted in the study. Pesticide-related illnesses are grossly underreported for a number of reasons: individuals may not seek or be able to afford medical care, doctors are often not trained to recognize pesticide-related illness, and symptoms of minor or even moderate pesticide poisoning can resemble those of other common ailments.

According to the study, children were the victims in 76 percent of the reported cases, and insecticides alone or combined with other pesticides were most often responsible for 895 cases (or 35 percent of the total incidents). Disinfectants caused 830 cases (32 percent), repellants were responsible for 335 incidents (13 percent), and herbicides were the cause in 279 cases (11 percent).

Organophosphates were the class of insecticides most frequently responsible for poisonings. Children are more vulnerable than adults to the effects of OP pesticides, which have been linked in animal studies to developmental delays, behavioral disorders, and motor dysfunction.

The study focused on acute pesticide exposure, but the authors expressed considerable concern for long-term effects: "Repeated pesticide applications on school grounds raise concerns about persistent low level exposures to pesticides at schools." The authors continued, "The chronic long-term impacts of pesticide exposures have not been comprehensively evaluated; therefore, the potential for chronic health effects from pesticide exposures at schools should not be dismissed. Unfortunately, the surveillance methods used in our report are inadequate for assessing chronic effects." The authors also noted that pesticides on school grounds can be tracked inside school buildings. Once inside, pesticides break down more slowly, with residues remaining for months or even years.[13]

Eight Fallacies About Pesticides

Fallacy #1: Pesticides prevent pests.

Fact: This is a classic misconception. Pesticides may kill pests, but they do not prevent them.

The IPM approach is successful because it is largely preventative. IPM involves simple approaches to exclude pests in the first place, combined with good sanitation and an awareness of what creates good pest habitats. Pesticides may be used in an IPM program, but are preferably of low toxicity and employed along with other measures.

Fallacy #2: Pesticides are not a health risk for children.

Fact: Pesticides are more of a health risk for children than adults and adolescents. A child's smaller body size and greater surface area per kilogram of body weight means it takes less pesticide exposure to affect his or her developing organs and tissues. Children's habit of playing on the floor or ground, exploring, and putting things in their mouths predisposes them to greater exposure than adults wherever pesticides are used.

Fallacy #3: Schools should not be burdened with providing the entire school community prior notification of pesticide applications.

Fact: Universal notification includes the entire school community, as opposed to a registry of people who ask to be notified. Universal notification ensures that staff, busy parents, and guardians are informed about pesticide applications in their child's school environment. It is by far the most efficient way of notifying the school community.

Fallacy #4: Parents and staff only need twenty-four hours' prior notification to the use of pesticides at schools.

Fact: It is the law in many states that schools must provide seventy-two hours' advanced notice to parents and staff prior to a pesticide application. Schools are required to pre-notify by posting signs in and around the school where the pesticide is to be applied. This includes baits, gels, and pastes that may fall under the EPA's definition of "least-toxic pesticide." Furthermore, normally only those holding an applicator's license are allowed to apply pesticides.

Teachers with a can of Raid under the sink for the odd pest emergency are not only breaking the law but are also putting their students' health at risk. Many products labeled for indoor crawling and flying pests contain pesticides that frequently trigger asthma attacks.

Fallacy #5: Pesticides are only harmful if they are sprayed directly on someone.

Fact: What we do know is that commonly used pesticides in schools today contain compounds which can take days or even weeks to wear off, particularly indoors or during the winter. Pesticide residues can enter the human body by accidental ingestion, skin absorption, or inhalation.

Fallacy #6: Pesticides are necessary because IPM is too expensive for schools and homes.

Fact: The initial cost associated with addressing long-overdue maintenance needs can be a hurdle the first year of an IPM program. Once maintenance issues have been addressed (such as door sweeps or general pest proofing) and the school community is practicing real IPM, the pest management costs go down and may continue to decrease.

Fallacy #7: Pesticides are tested extensively before being approved. If they weren't safe, the government wouldn't let us have them.

Fact: For a number of reasons, current testing methods do not fully address pesticide toxicity in humans, and no standard testing method is conducted that would predict and warn against health effects to those most vulnerable, children. Even when it comes to adults, the long-term health effects of many pesticides already in use are unknown. Though the law requires that any application of pesticides on school grounds be done by a licensed professional, they nevertheless may pose risks to adults, adolescents, and individuals with health problems and especially children in ways science does not yet fully understand.

Fallacy #8: More is better.

Fact: Actually, more can be deadly. Labels are very carefully written and directions should be followed to the letter.[14]

IPM in Schools

Use of IPM principles and practices in school environments is a growing trend in communities throughout the United States. IPM focuses on pest prevention using effective, least-toxic methods, and is proving practical to apply and cost-effective to operate.

A school is a challenging place to operate a pest management program. Most school buildings are unintentionally designed with ideal entry points and locations for pest insects, rodents, and other unwelcome wildlife. Inappropriate landscape design and plant selection often encourage weeds and other pest problems. Diminishing budgets and deferred maintenance exacerbate these conditions.

Schools also include diverse physical spaces, both indoors and out, that require customized solutions to pest problems. In addition, schools host a wide variety of people, from teachers and students to vendors and community groups, who have differing opinions about pest tolerance levels and appropriate pest management methods. It is necessary to sensitively address the concerns of parents and others who desire a school site free of nuisance or health-threatening pests, but want this achieved with minimal use of toxic materials.

Because IPM is a decision-making process and not a rote method, an IPM program will always be able to take into account the wide spectrum of pest problems and the diversity of people involved. IPM methods enable pest control operators and other members of the IPM team to design flexible, site-specific pest management plans scaled to the severity of the problem and the level of resources available.

The IPM approach also offers unique opportunities to incorporate pest management issues into the school science curriculum and gives students hands-on learning experiences in the biology, ecology, and least-toxic management of pests that inhabit school buildings and grounds.

In the IPM approach, considerable effort is also put toward preventing pest problems by controlling conditions that may attract and support pests. For example, to control an infestation of pavement ants in a classroom, placing ant baits (pesticides) in areas inaccessible to children, or applying gel baits to baseboards might be effective options. However, many schools have experienced repeat problems because pesticides alone do not usually achieve long-term control. Some schools have also received complaints when repeated, frequent applications of a pesticide occur in areas where children are present every day.

For long-term pest control, it is essential to identify why the infestation arose in the first place. Non-chemical controls such as sealing cracks and crevices to prevent access, improving sanitation around food preparation and waste/recycling storage areas, and limiting where food can be eaten will help prevent many problems. IPM addresses the cause of the problem (food scraps and crumbs throughout the building) to avoid the development of the symptoms of the ant infestation.[15]

Educating IPM Participants

A school IPM program should include a commitment to the education of students, staff, and parents. This education should include not just teachers, but also school nurses, cafeteria workers, housekeeping staff, and administrative personnel as well. All school occupants must understand the basic concepts of IPM and who to contact with questions or problems. Specific instructions should be provided on what to do and what not to do.

Faculty and staff work in the classrooms and cafeteria areas just like students. As a result, teachers and staff are exposed to all of the same risks as students. In addition, faculty and staff should not introduce potentially harmful bug sprays into the classroom. Commonly used over-the-counter products available at local stores often contain the same ingredients as those products available only to licensed pest control operators. When used in the classroom, these sprays are potentially dangerous to chemically sensitive children. Also, these products can make some pest problems worse because they may interfere with or even reduce the effectiveness of treatments made previously by the pest management staff.

Many schools across the United States have incorporated environmental issues into their curricula. Science classes might include discussions and activities to learn more about the fascinating and diverse roles of insects, plants, rodents, and birds in our world. Most are harmless, and many, for example, some spiders, predatory mites, centipedes, and certain beetles, are actually beneficial in controlling pest populations. If good sanitation is practiced, the population of these beneficial insects can be kept at tolerable levels.

Staff and students need to understand how their own behavior helps alleviate or contributes to pest problems. School staff should encourage the parent-teacher association, student organizations, and other school-affiliated groups to participate in the IPM program.

The first step to reducing pesticide use is to understand the risks of pesticide exposure. If the problem can be laid out clearly in a school district, including listing what pesticides are used and their health effects, it will be a powerful tool in calling for change. Pesticide use and pest management policies vary dramatically from school district to school district. Determining how pest management decisions are made and what types of pesticides are used is the place to start.

Informational Roadblocks

Unfortunately, finding the answers is not as simple as asking the right questions. In fact, the process may be long and frustrating. There are several reasons for this. The first and most fundamental problem in many school districts is that there is no one person responsible for pest management within the district. Because of this, it often takes many calls to simply determine who the responsible person is.

Second, most school districts have no pest management policy, often resulting in very poor record keeping. It is fairly common that no one person can indicate what pesticides are being used in the school district because no one person is responsible for tracking that information. Third, school districts today are faced with many problems, and reducing pesticide use often is a low-priority item despite its potential health impacts. Finally, many school districts are resistant to provide this information because they know that it may be used by community activists calling for change. In many school districts, lack of knowledge about the health impacts of pesticide use on children and the viability of alternatives leads them to oppose change.[16]

Costs of IPM and Conventional Programs

A major impediment to the adoption of IPM practices in schools is the perception that they incur higher costs. Indeed, the start-up costs of an IPM program may be higher than the costs associated with an ongoing conventional pest management program. However, a conventional method also incurs high initial costs, suggesting that initial higher costs of both programs may be related to the expenditure of more time becoming familiar with the elements of each type of program.

Nevertheless, there are several expected cost advantages to the IPM approach that may be overlooked. Labor, and thus the duration of each service, is the major contributor to overall cost.

Whereas most IPM-related tasks (for example, caulking and baiting) can be done during school hours, resulting in more flexible work time, conventional services (such as baseboard and crack-and-crevice spraying) require that all people vacate the rooms. In conventionally serviced schools, the pest management professional (PMP) routinely waits for students to be dismissed before initiating a pesticide application. More importantly, some pesticides, primarily baits used in IPM programs, have long residual activity and are generally placed in areas that are less likely to be exposed to routine cleaning. Therefore, over the long run, it is expected that subsequent services would use less bait, resulting in cost savings in materials and further reducing pesticide exposure to occupants.

However, cost estimates of IPM services do not include time spent on training the PMPs because they had received earlier training in general pest control and require only a brief refresher in IPM techniques. Because school personnel in many districts are responsible for pest control services and they may not be familiar with IPM, much more extensive training would be necessary for proficiency in IPM tactics. Consequently, as certification requirements change and IPM becomes a common element of PMP training, training costs are expected to be no different than for conventional pest control, and no cost adjustment would be necessary. The shift to IPM will obviously accelerate if schools specify in their pest control contracts that only individuals trained in IPM may furnish services.

Compare the total costs of a conventional pest management program with the costs of an IPM program. Instead of routinely spraying pesticides, IPM employs monitoring to determine the location, extent, and the cause of a weed or pest problem, and then applies a variety of non-chemical or least-toxic pesticide controls. IPM strategies are more effective because they are better able to prevent pest troubles. This is because they focus on modifying the cause of the dilemma, instead of just spraying the pest after it becomes a problem. IPM is the equivalent of a wellness program to maintain the facility and landscape in a healthy condition to avoid the need for specific chemical treatments and the costly side effects that can be associated with them. Pesticides are used only after other options have been fully considered and only if other methods have not reduced pests to tolerable levels. Determination of this tolerance level is based on pest-specific and site-specific criteria.[17]

Least-Toxic IPM

Least-toxic IPM decision making seeks to manage pests through prevention. It is based on the fact that pests almost always can be managed without toxic chemicals. Prevention is the first line of defense. Modification of pest habitats (such as putting vegetation-free buffer zones alongside buildings) deters pests and minimizes infestation. IPM requires extensive knowledge about pests, such as information about infestation thresholds, life cycles, environmental considerations, and natural enemies. Pest monitoring is critical to identify existing pest problems and areas of potential concern, as well as to determine how decisions and practices may impact future pest populations. Monitoring must be ongoing to prevent small pest problems easily controlled with least-toxic means from becoming infestations. Threshold tolerance levels of pest populations are established to guide decisions about when pests pose a problem sufficient to warrant some level of treatment. If treatment is necessary, non-chemical means are given priority. Traps and enclosed baits, beneficial organisms, freezing and flame or heat treatments, among others, are all examples of non-chemical or least-toxic pest treatment methods. A good IPM program prohibits the use of known and probable carcinogens, reproductive or developmental toxins, endocrine disrupters, cholinesterase-inhibiting nerve toxins, and the most acutely toxic pesticides.

In sum, least-toxic IPM establishes a hierarchy of appropriate pest management strategies, with monitoring and prevention at the top and toxic pesticides at the bottom. Least-toxic IPM never provides all available pest control methods equal consideration. It always favors non-toxic alternatives. Beware of alleged IPM policies that allow the use of chemical pesticides without prior exhaustion of all other means of control.[18]

Least-Toxic Approaches Are Cost-effective

Preliminary indications from IPM programs in school systems suggest that long-term costs of IPM may be less than those of conventional pest control methods. By focusing on prevention and monitoring whenever pests are a problem, school IPM programs may require no treatments at all. Short-term costs may include IPM training, new equipment purchases, hiring an IPM coordinator, or preliminary school building repairs. However, in contrast with chemical-intensive methods, over the long term IPM garners savings by eliminating or reducing ongoing chemical purchases and applications.[19]

Everyone agrees that a good IPM program effectively controls pests. According to the EPA, "preliminary indications from IPM programs in school systems suggest that long-term costs of IPM may be less than a conventional pest control program."[20] Schools around the country have saved money using IPM methods.

However, costs should not be the most important consideration in evaluating the success of IPM programs. The incalculable benefit of a healthier environment for our children should be the dominant motivation in the minds of decision-makers.

Examples of IPM Effectiveness and Cost Efficiency

Public schools in Montgomery County, Maryland, produced cost savings and realized better pest control when they shifted from a traditional spray program to IPM for their buildings and landscapes. A crude comparison of labor, equipment, and materials costs showed savings ranging from 15 to 18 percent per year over a six-year period. Pest control costs were reduced by $111,000. The school district saved $1,800 at each school and $30,000 at its food service warehouse. In Montgomery County, reduction of school pesticide use by 90 percent and use of least-toxic pesticides when pesticides are required have made school and work safer for 110,000 students and 12,000 employees.[21]

In Monroe County, Indiana, a school IPM program decreased pest management costs by $6,000 in two years. Pesticide use reportedly plummeted 90 percent, and all aerosol and liquid pesticides were discontinued.[22] The IPM specialist in Monroe County stated that "costs are dependent on the condition of the school. We were lucky in this case that Monroe County began this project with a history of good management practices and structurally sound school buildings. If a school isn't in good shape maintenance wise, the startup costs of an IPM program can be a little higher in the beginning."[23]

In New York state, after Susquehanna schoolchildren were accidentally exposed to pesticides and became ill in 1991, the New York Department of Environmental Conservation ordered the school to halt all routine pesticide applications and to implement an IPM program. Engineers and the pest control company on contract are now pleased with improvements in the indoor environment. Prior to the IPM program, the school was sprayed monthly for recurring ant problems. Now with monitoring, increased sanitation, education, and the use of least-toxic baits only when needed, the number of ant sightings has decreased substantially while costs have declined.

Pesticide applications have been discontinued altogether in Susquehanna's outdoors environment as the school's engineers manage the turf and playing fields. They now use organic fertilizer and compost twice a year on the athletic fields, aerate the soil four times a year, and mow high and often. An engineer at the school says, "Cost will depend somewhat on how much labor you need to get the job done. In our case, we spent the first years doing some preventative maintenance such as putting plastic lining under the bleachers and deeply aerating the fields. We have now cut costs and labor across the board for the past seven years and the turf looks better than ever."[24] The Susquehanna school is currently saving more than $1,000 annually on pest control with its new IPM program.

Several school districts in North Carolina are using IPM to lower the frequency of pest infestations as well as the cost of pest control. During the summer of 2003, the Agricultural Resources Center & Pesticide Education Project surveyed the facilities departments at all of North Carolina's 117 public school districts regarding their pest management practices. Sixty districts responded, representing more than half of North Carolina school districts and 1.3 million students from kindergarten through grade twelve. The survey found that many schools still use high-risk pest control practices such as fogging buildings with pesticides or using pesticides regularly as prevention. Schools with least-toxic or IPM programs consistently spent less than the statewide average on pest control, and tended to be more satisfied with their pest management programs overall. Some of the survey's most interesting findings include: 1) on average, North Carolina school districts spend $1.77 per student per year on pest control, whereas districts with least-toxic pest control programs, such as IPM, spend $1.49 per student per year; 2) 43 percent of school districts report using pesticides regularly in classrooms; 3) 17 percent of school districts fog buildings with pesticides; 4) only three school districts reported notifying parents when pesticides are used at school; 5) 65 percent of districts report consciously selecting least-toxic pesticide product formulations; 6) large urban districts as well as small rural districts in North Carolina report success with IPM programs.[25]

IPM's cluster of preventive approaches (cultural, mechanical, and biological) are easy to implement because they can be incorporated into schools' existing custodial and maintenance functions, such as sanitation, energy conservation, building security, and infrastructure maintenance.

"In-House" or Contracted Services

IPM programs can be successfully implemented by "in-house" school employees or by contracting with a pest control company. A combination of in-house services with contracted functions may be mixed and matched to the needs and capabilities of each school system. Both approaches have advantages and disadvantages. Individual school systems must decide what is best for them given their unique circumstances. Whether selecting in-house or contracted services, pest management personnel should be trained to: 1) understand the principles of IPM; 2) identify pests and associated problems or damage; 3) monitor infestation levels and keep records; 4) know cultural or alternative methods; 5) follow recommended methods of judicious pesticide application; 6) learn the hazards of pesticides and the safety precautions to be taken; and 7) understand the pesticide label's precautionary statement(s) pertaining to exposure to humans or animals.[26]

Status of State Pesticide Laws

Federal legislation to require safer pest control in schools has been stalled in the U.S. House of Representatives for several years. Introduced in 1999 and adopted twice by the Senate, the School Environmental Protection Act requires schools to adopt less-toxic methods of pest control and notify parents and staff when pesticides are applied on school grounds. The legislation is the product of years of effort by parent coalitions working to bring safer pest control practices to their schools.

FIFRA currently requires the registration and review of all pesticides produced and distributed in the United States. But it does not address school applications. The FQPA of 1996 amended relevant sections of FIFRA but simply defines IPM to include chemical agents when necessary.

In the absence of a federal mandate, state policies on pesticide use in schools are widely disparate and provide mixed protections. Texas and Michigan were the first two states to enact legislation that required schools to follow school IPM pest management plans. Since the late 1980s and early 1990s, thirty-four states have jumped on the bandwagon, creating a patchwork of different school pest management programs. Kentucky, Maine, New Jersey, Ohio, Pennsylvania, and Rhode Island either approved or enacted their own requirements for IPM in schools. In 2003, even more states considered school pest management plans. These states included Utah, Indiana, Arizona, and Illinois, which is expanding its program. Alaska, Illinois, Pennsylvania, and Washington state require that schools notify everyone in the school system the beginning of the school year about the pesticides used at their school. Baits, gels, pastes, and antimicrobials are exempt from this notification requirement. Twenty-nine states require that schools post signs stating that pesticides will be sprayed; these signs must be posted twenty-four to seventy-two hours before spraying. Sixteen states require notification of indoor applications. Twenty-six states require postings outdoors. Thirteen states require more information beyond label requirements. Many states have reentry requirements, wherein pesticide use is completely prohibited while school is in session. Massachusetts orders that no pesticides be sprayed in schools without permission. Nineteen states approved IPM legislation or rules that will require schools to define and submit a written pest management plan to a regulatory agency. Many states have voluntary IPM requirements in schools; these states include California, Connecticut, Minnesota, Montana, New York, and Utah. Other states have mandatory requirements for IPM in schools; these states include Florida, Illinois, Kentucky, Louisiana, Maryland, Massachusetts, Maine, Michigan, New Jersey, Pennsylvania, Rhode Island, Texas, and West Virginia. Sixteen states have no laws concerning pesticide use in schools.[27] Eight states restrict the aerial or large broadcast application of pesticides in areas neighboring schools. These buffer zones can range from thirty-five feet to 2.5 miles.

Buffer zones around schools could impact spraying initiatives like those to combat virus-bearing mosquitoes. In an effort to prevent windborne drifting of potentially harmful chemicals, states such as Alabama, Arizona, Louisiana, New Hampshire, New Jersey, and North Carolina limit the use of pesticides within a specified radius of school buildings.

But these laws are hardly uniform. Alabama outlaws aerial spraying or dusting within 400 feet of school grounds. Louisiana broadens the circle to 1,000 feet but lifts the restriction before and after school hours. New Jersey's buffer zone provision, by far the most complex, includes a separate set of parameters for gypsy moth chemicals and has different zone measurements for grade schools and high schools.[28] Illinois's law, passed in 1999, protects children from pesticides. It was one of the first of its kind calling for IPM. However, although the law has been on the books for nearly a decade, many schools are still not in compliance due to the lack of state-sponsored IPM education programs for school administrators. In order to protect their children, parents must ask their schools to comply fully with regulations in their state.[29] Massachusetts law, as of November 1, 2000, requires that schools and daycare centers maintain records of all pesticide applications for at least five years. In addition, due to the persistence of pesticides used for termite control, records of pesticide applications for termite control should be kept on record for the life of the property.[30]

Non-Compliance in Massachusetts

A recent report concludes that most parents in Massachusetts are not being informed of pesticide use at their children's schools and childcare centers despite a state law requiring schools to submit detailed plans about pesticide use on their grounds. The Children's Protection Act of 2000 requires schools and daycare centers to submit plans explaining the extent of their pest problems, the pesticides they plan to use, and who will apply them. The law also requires parents and teachers to be notified at least two days before spraying begins. The report found 81 percent of the state's schools and childcare centers have failed to comply with the law. According to the audit, more than 70 percent of the state's 2,456 schools, both public and private, and 90 percent of the state's 3,242 childcare centers have not submitted the required pesticide plans.[31]

The California Statute

After activists waged a three-year campaign to highlight the widespread threats to children's health, Governor Gray Davis signed the Healthy Schools Act into law in September 2000. The law requires school districts to: notify parents annually about what pesticides the district intends to use in their children's schools and on school grounds during the following year; provide parents the option to register to be notified seventy-two hours in advance of all pesticide applications and seventy-two hours after application; and maintain records of all pesticide use for four years in an accessible format.

The law also requires the state Department of Pesticide Regulation to: provide training for interested school district personnel in least-toxic Integrated Pest Management techniques; distribute a manual to all schools in least-toxic Integrated Pest Management; and maintain a Web site with information to help schools comply with the law and implement least-toxic pest management methods.

Impact of the Healthy Schools Act

Although many of California's largest school districts have moved to reduce the use of dangerous pesticides since passage of the Healthy Schools Act, fifty-four active pesticide ingredients that are known or suspected carcinogens, reproductive or developmental toxins, endocrine disrupters, or acute toxins and/or cholinesterase inhibitors may still be in use in and around California schools. This is twelve more active ingredients than districts reported using in 1999.

Furthermore, a re-survey found that by late January 2002—one year after the Healthy Schools Act went into effect and almost six months after the first full school year under the act began—one-third of California school districts were not in compliance with the act's parental notification requirements. This deficiency deprives parents of important information regarding their children's safety and health.

Moreover, many school districts that complied with notification requirements were still unable or unwilling to produce records concerning pesticide use and application. The ease of finding out which and how frequently pesticides are applied and how many parents are registered for notification before each application varied greatly among California's school districts. For example, the Long Beach Unified district returned the survey almost blank, while the Elk Grove Unified and San Juan Unified districts required nearly two months of follow-up calls to return even the most basic elements of requested information.[32]

One need not be a rocket scientist to reach the undeniable conclusion that despite a ballyhooed law, dangerous pesticides use continues to threaten children's health in California schools.

Pesticide Regulation in Ohio

In Ohio, the law requires only that commercial applicators post signs when a lawn pesticide has been applied, be it an herbicide, insecticide, or fungicide. Some Ohio school systems have adopted practices that reflect concern about children's exposure to toxic substances. Yet, a survey found that Ohio schools use very poisonous chemicals to control mere nuisance pests such as ants without warning students and parents when pesticides are applied, and they are relatively careless about when the pesticides are applied.

In May 2003, an exterminator sprayed weed killer around Madison Middle School in Madison, Ohio. As a result, fresh spring air pouring into the sixth- and seventh-grade classroom windows became tainted with the acrid smell of Formula 190, sending one teacher home and forty-two nauseated and dizzy children to the hospital. A new pesticide agent has been hired, all work must occur on weekends, and indoor spraying must occur in summer. Although written standards of pest control have not been yet adopted, all work must be scheduled through the maintenance supervisor.[33]

The Washington State Law

School districts throughout the state of Washington routinely use pesticides linked to cancer, nervous system damage, reproductive harm, and hormone disruption.

These are called high-hazard pesticides. In a 1998 survey by the Washington Toxics Coalition, 88 percent of thirty-three school districts reported using at least one high-hazard pesticide. School districts surveyed represented a range of rural, urban, small, and large districts, so the hazards of school pesticide use appear to be widespread.

Use of pesticides that can cause serious health effects faced no special restrictions in Washington schools unless an individual school district took action to protect its students and staff. School districts were not required to automatically notify all parents or compile yearly reports of pesticide use. A request for information about pesticide use might yield no response or a huge stack of application records.

As had occurred in California, activists achieved success after waging a five-year campaign to highlight the widespread threat to children's health. Governor Gary Locke signed the Children's Pesticide Right-to-Know Act into law in May of 2001. The law required school districts to: notify parents annually about their pest management policies and methods, including posting and notification requirements; maintain records of all pesticide applications to school facilities and make records readily accessible; provide an annual summary of all pesticide use in the district during the previous year; notify at least interested parents or all parents forty-eight hours in advance of all pesticide applications, for example, with a registry; post notices forty-eight hours in advance in a prominent place in the main office.[34]

Restricted Spray Zones Around School Property—Recommendations

Pesticides move off the target site when they are sprayed, whether indoors or outdoors. When sprayed outside, pesticides drift onto nearby properties, resulting in off-target residues. In order to adequately protect against drift, buffer zones should, at a minimum, be established in a two-mile radius around the school's property. Aerial applications should have larger buffer zones, at least three miles encircling the school. Buffer zones should be in effect at all times of day. It is especially important for spray restrictions to be in place during commuting times and while students and employees are on school grounds.

Posting Notification Signs for Indoor Pesticide Applications—Recommendations

States employ different approaches in providing school pesticide use information to parents, students, and staff. Some forms include posting notification signs and/or distributing notices directly to the affected population. This is a vehicle for a basic right-to-know if the notices are posted where parents, students, and staff can easily see them. It is important to post signs for indoor pesticide applications because of the extensive period of time students and school employees spend at school. Signs posted prior to commencement of the pesticide application, not after, are more protective because they effectively enable people to take precautionary action. Because of the residues left behind after an application, signs should remain posted for at least seventy-two hours. It takes time for pesticides to break down and some pesticide residues can last for weeks. Signs should also be posted at all main entrances of the building and the specific area sprayed, on the main bulletin board, and, for more comprehensive notification, in the school newspaper or on the daily announcements. Posted signs should state when and where a pesticide is applied, the name of the pesticide applied, and how to get further information, such as a copy of the material safety data sheet (MSDS) and the product(s) label.

Posting Notification Signs for Outdoor Pesticide Applications—Recommendations

For a wider range of protection, states should require posting pesticide notification signs for outdoor pesticide application as well as indoor spraying. Students who play sports or people continually on the lawns are at greater risk when applications occur on school property. Dermal exposure can occur when a football player gets tackled, a soccer player slides to make a block, or a student sits on the grass eating lunch or watching a game. Inhalation exposure can occur when a player breathes in kicked-up dust and dirt and pesticide residues. Even spectators at a game or passersby face inhalation exposure to pesticides that volatilize or vaporize off the treated area.

Prohibition on Use—Recommendations

Limiting when and what pesticides are applied in and around schools is important to the reduction of pesticide exposure. Pesticides should never be applied when students or employees are in the area or may be in the area within twenty-four hours of the application. In reality, certain types of pesticides, such as carcinogens, endocrine disrupters, reproductive toxins, developmental toxins, neurotoxins, and persistent compounds should not be used around children.[35]

Sample IPM Notification Letter

A standard IPM notification letter to school districts explains the following details: Dear Parent(s) or Guardian(s):

The Peters Township School District uses an Integrated Pest Management (IPM) approach for managing insects, rodents, and weeds. Our goal is to protect every student from pesticide exposure by using an IPM approach to pest management. Our IPM approach focuses on making the school buildings and grounds an unfavorable habitat for these pests by removing food and water sources and eliminating their hiding and breeding places. We accomplish this through routine cleaning and maintenance. We routinely monitor the school building and grounds to detect any pests that are present. The pest monitoring team consists of our maintenance personnel and building staff. Pest sightings are reported to our IPM coordinator, who evaluates the "pest problem" and determines the appropriate pest management techniques to address the problem. The techniques can include increased sanitation, modifying storage practices, sealing entry points, physically removing the pest, etc.

From time to time, it may be necessary to use chemicals to manage a pest problem. Chemicals will only be used when necessary, and will not be routinely applied. When chemicals are used, the school will try to use the least-toxic product when possible. (Applications of chemicals will be made only after normal school hours.) Notices will be posted in these areas seventy-two hours prior to application and for two days following the application.

Parents or guardians of students enrolled in the school may request prior notification of pesticide applications made at the school. To receive notification, you must be placed on the school notification registry. If you would like to be placed on the registry, please notify the district in writing.

If a chemical application must be made to control an emergency pest problem (for example, stinging insects), notice will be provided by telephone to any parent or guardian who requested such notification in writing. Exemptions to this notification include disinfectants and anti-microbial products; self-containerized baits placed in areas not accessible to students; gel-type baits placed in cracks, crevices, or voids; and swimming pool maintenance chemicals.

Each year the district will prepare a new notification registry.

If you have any questions, please contact Bruce Riebel, IPM Coordinator[36]

While notification alone will not protect children's health, it provides important information and can be a great tool to advocate for a least-toxic pest management policy.

IPM Responsibilities: A United Effort

Kitchen Staff

Food handling and preparation areas are among the most crucial areas for pest management. It is imperative that kitchen staff understand the importance of good sanitation, kitchen management, and proper food storage. For example, lids should be kept on garbage cans, spills cleaned as soon as possible, and food stored in pest-proof containers. A well-trained kitchen staff can assist the district's IPM staff in locating and eliminating pest harborage areas. Kitchen staff should also be involved in periodic IPM training.

Administrators

Administrators and school boards set the tone for an IPM program. Their first responsibilities are selecting a qualified individual for the IPM coordinator's position and establishing a pest management policy. Administrators should have a general understanding of suggestions pertaining to IPM in schools, the possible penalties for improper pesticide use by in-house pesticide applicators, and pesticide safety issues and decision making about which pesticide products are appropriate for district use.

Perhaps the most crucial role of administration is assigning priorities for building maintenance requests submitted by the IPM coordinator. Without administrative support for such requests, as well as requests to correct other reported problems (such as inadequate sanitation or improper management practices), IPM programs will be restricted in their effectiveness.

Teachers and Students

In addition to food handling areas, classrooms and lockers are key sites for pest problems in school buildings. The most important responsibility of the students and faculty is sanitation: cleaning up leftover food, proper storage of pet food and snacks, and maintaining uncluttered and clean classrooms and locker areas. Also, teachers and students who can identify pests can be helpful to the IPM program. The more participants, the greater the potential for success. Without the cooperation of teachers and students in the IPM program by reporting pests immediately and by keeping their classrooms clean, it is probably unreasonable to expect a totally pest-free environment or to control pests without any use of pesticides.

Parents and Community

Parents want their children to experience a pleasant learning environment without any undue risks from pesticides. For this reason, parents are usually among the first to express their concerns about perceived unsafe conditions in a school. Unsafe conditions can occur when pest problems are improperly managed, or when pesticides are overused or used improperly. Parents should be made aware of the current pest management practices in their children's schools. Visible interest and concern on the parents' part can stimulate the school to do its best to provide effective and safe pest control in school facilities. Parents and other community members can and should express their views to the IPM coordinator, school superintendent, school board, campus-based improvement committees, and parent-teacher associations and organizations. Parents can serve on IPM advisory committees with the schools.

Community involvement is crucial in the development and success of school pesticide-reduction programs. Nearly all of the schools that have reduced or eliminated the use of pesticides on school grounds have done so because of parent and community pressure. Parents, teachers, students, and community members have advocated for healthy classrooms and playgrounds. These local groups have worked with grounds and maintenance crews, built public support, obtained coverage in newspapers and on television, and effectively lobbied school board members.[37]

The Role of the School Nurse

The school nurse is the health expert in the school building. Training and experience in physiology, chemistry, biology, and health care can position the school nurse to be actively involved in addressing issues of toxin exposure. Nurses are important to environmental health because they can play key roles in protecting the health of all people; are in contact with staff, students, families, and community members; and have credibility and access that enables them to provide scientifically sound information about environmental issues and exposures.[38]

Of specific importance to school nurses is the impact of environmental exposure for the growing child. Schools have programs serving young infants, toddlers, and pre-schoolers, as well as traditional school-age children and youth. The child's metabolism, developing body systems, and environmental components can interact in ways similar to and different from those of adults. Also, some exposures, while not apparently harmful to adults in similar doses, can result in adverse health effects for children. Childhood exposures may result in health problems years later. Children differ from adults in the exposure dose levels, routes, response potentials, and response effects.[39]

The school nurse is concerned with child toxin exposures that have occurred prior to school attendance and have impaired a child's health or learning potential (such as lead exposure causing neurological and learning deficits). Additionally, the nurse is concerned with preventing adverse health reactions in children exposed to physical, chemical, and biological toxins occurring in the school environment.[40]

The Occupational Health and Safety Administration (OSHA) has many regulations that impact the safety of school workers, especially those relating to occupational exposure to environmental toxins. Safe practices for employees translate into safer practices and environments for children. The school nurse can assist in implementing safety standards and in making the standards understandable and applicable to the school setting.[41]

However, the nurse's role does not stop with treating or evaluating existing health hazards to children. She must also take a proactive role to anticipate health hazards and act to reduce or prevent adverse health effects on the school population. For example, the nurse may note that pesticides are routinely sprayed during the morning hours every three months. Some who have been exposed may seek her help. Her proactive role here would be to work with the school administration and the pesticide applicator to spray on Friday afternoons after school, so that when students return on Monday morning, there are much lower amounts of pesticide residues in the school and thus much less of a potential health impact.

Parents' Pesticide Awareness

While parents of school-age children may be deeply concerned about adverse health effects in the school environment, their awareness of the threat posed by pesticides may be inadequate.

The fact of the matter is that most parents do not often think about pests or the use of pesticides in their children's schools, but when these issues are raised, they strike a deep emotional chord. That is one of the findings from focus group discussions with parents.

When asked which environmental quality issues in their children's schools concerned them most, parents mentioned air quality, asbestos, and lead and mold/fungus problems, but not pests or pesticides. While pesticides are clearly not top concerns for parents, this research shows they do become easily concerned when the subject of pesticides is raised.

This qualitative research was conducted in April 2001 among parents of kindergarten and elementary-school children in four cities: Cambridge, Massachusetts; Raleigh, North Carolina; Chicago; and Los Angeles. The focus groups were comprised of low- to moderate-income parents and those with children in urban school settings. All of the parents were actively involved in their children's school-based activities. This research helped to understand parents' perceptions and attitudes about pests and the use of pest control products in and around schools.

While pesticide use in schools is not a top concern for parents, neither is the health and safety threats posed by pests. While parents have an aversion to pests, primarily because they are viewed as an indication of unsanitary conditions, they know surprisingly little about the actual health hazards associated with even the most common pests.

Other findings from the parent focus group discussions include: Rats and roaches are perceived by parents as the worst pests because they carry diseases (non-specific), promote unsanitary conditions (droppings), and pose a general threat to a child's well-being. Lacking knowledge and information about pesticides, parents default to erroneous misperceptions about pesticides, such as: all pesticides are sprayed; pesticides used in schools are industrial strength; school janitors are responsible for applying pesticides; and applications are widespread (the entire school building) and random (not part of a strategy or plan).

Parents know little about IPM, but when given its legal definition, they see it as a logical, commonsense approach to controlling pests. When asked about pre-notification of pesticide applications in schools, most parents want a general understanding of what's being done, but are not interested in knowing about every application.[42]

School Pesticide Incidents Across the Nation

January 1999, Mandeville, Louisiana. Two students at Mandeville Junior High School were exposed to Green Thumb Wasp and Hornet Killer (tetramethrin and phenothrin) sprayed in an effort to kill ants by a teacher in a practice room for the school band. One parent filed a health complaint expressing concern about possible health effects, though no symptoms were reported at the time. The spraying was reportedly done in violation of the district's pest management program. The state issued a warning letter to the district citing several violations of the state pesticide law, including applying a pesticide in a manner inconsistent with its labeling, allowing a person to apply pesticides who was not a certified applicator, not keeping a record of the application, and applying a pesticide in a school while children were present or expected to be present within eight hours.[43]

November 25, 1998, Washington County, New York. A parent filed a complaint with the state after seeing a school custodian at Greenwich Central School with Claire Lice Killer (pyrethrins and piperonyl butoxide) and being told that the chemical was applied in classrooms in an effort to control head lice. The parent's complaint also noted that she had seen school staff applying diazinon to control bees near the school building, and that her son had seen a janitor spraying playground equipment. The parent was concerned about potential exposure to her son and other students. An investigation showed that the school nurse had requested the spraying for head lice. Following an investigation, the district was cited for multiple violations of state pesticide law, including use of a pesticide product not registered in the state, authorizing applications by uncertified employees, failure to keep records of applications, and failure to post notices at athletic fields that had been treated with Roundup (which contains glyphosate). The school signed a consent order and a $2,500 penalty was suspended.[44]

November 12, 1998, Mount Pleasant, South Carolina. A pest control firm mistakenly drilled through wall voids and into two classroom walls at Laing Middle School during a "trench-and-rod" termiticide application to the building's exterior foundation. The pesticide Dursban TC (chlorpyrifos), which is not registered for interior use, was injected into the holes and into at least one of the classrooms. The teacher reported a strong odor upon returning to the classroom the next morning. She reported it to the principal, and the room was aired out. When the odor remained the following week, the students were moved to another classroom, and the pest control applicator was called back to the school. He noted the strong smell at that time, patched the holes in the walls, and hired a company to clean the carpets, walls, ceiling, desks, and pencils in both classrooms. Some textbooks that had been contaminated with the pesticide were replaced. However, the odor persisted, and a second carpet cleaning and general cleanup was done in December.

A student mentioned the pesticide "spill" to a parent in late January, two and a half months after it had occurred. This parent talked with school staff and realized that the incident had not been reported to state agencies. She reported the incident, and only then were other parents notified. Parents began to wonder if strange illnesses their children had been experiencing, including flu-like symptoms and one child with peeling hands, may have been caused by exposure to the chemical. Chlorpyrifos residues were found in carpet samples collected by state investigators two and a half months after the application, after two professional carpet cleanings. The pest control company was cited and fined by the state for applying a pesticide in a manner inconsistent with its labeling. The school board later sued the pest control company.[45]

July 1998, Somerset, Wisconsin. Staff at the state Department of Agriculture, Trade, and Consumer Protection circulated a survey to state school districts inquiring about their pesticide use practices. Agency officials noted that St. Anne's School filled out the survey indicating that the school used chlordane, a persistent organochlorine pesticide that has been banned since 1988. An agency inspector visited the school and confiscated a partially used one-pound container of the pesticide. He was told that the product had been at the school since at least the start of the school year, and that it had been used once during the most recent school year. The state agency did some testing of sites where the chemical had been used, but no residues were found. No injury, illness, or exposure to the pesticide was alleged. However, as the chemical is classified by the EPA as a probable human carcinogen, a warning notice was issued.[46]

June 9, 1997, Chardon, Ohio. Seven staff members and nine students at Chardon High School experienced dizziness and difficulty breathing and were treated at local hospitals after fumes of the herbicide HNS-300 (bromacil) seeped into the building. It was applied by school maintenance workers in spot applications to the perimeter of the building and under stadium bleachers. The fumes were drawn into the school by the ventilation system, and came in through the gymnasium's windows. Complaints about odor were reported approximately one hour after the application was completed. The school was evacuated. The incident was not reported to the Ohio Department of Agriculture, although the agency did a site inspection three weeks later after learning about the incident through newspaper accounts. It was found that the school district did not have any licensed applicators on its staff. The state inspector issued a "field notice of warning," but no citations or fines were levied.[47]

March 20, 1997, Amherst, Massachusetts. A kindergarten class was moved out of Fort River Elementary School after complaints of persistent headaches from three staff members and a number of students throughout the spring. One teacher was out sick for weeks. The school had a history of indoor air quality problems. However, some people in the school suspected that applications of an ant spray, Double Active Residual (propoxur, pyrethrins, and piperonyl butoxide) might be contributing to the problem. Custodians had applied it multiple times inside the kindergarten classrooms in March.

The teachers' union filed a complaint with the Massachusetts Pesticide Bureau on June 9. The state inspected the school on June 19, three months after the pesticide applications were made and health symptoms began. Because of the untimely nature of the complaint, state investigators undertook no human or environmental testing. The state investigator concluded that the pesticide applications were made according to product labels, and that any symptoms that had occurred were unlikely to have been caused by the pesticides. This conclusion was reached based on the fact that the last application was made more than two months before the complaint was filed, and that the active ingredient of the product was pyrethrin at a very low concentration. (In fact, the product also contained another active ingredient, propoxur, at ten times the concentration of the pyrethrins. The EPA classifies propoxur as "moderately persistent." Incident information reported to the EPA between 1992 and 1996 showed that symptoms people experienced from post-application exposures to propoxur included headaches, nausea, and respiratory irritation.) The school principal was quoted in a newspaper account the day after school's evacuation as saying that exposure symptoms listed on the spray can are similar to ones that people reported experiencing. State investigators did note in their report that the school employees who made the applications were not licensed, and that notices of the applications had not been posted as required by state law.[48]

August 1994, Pierre Part, Louisiana. In the week before school opened for the fall, a school custodian sprayed the schoolyard around Pierre Part Primary School with the unregistered insecticide lindane in an effort to control rodents and fleas. Diazinon was also sprayed in and around fourteen portable classrooms just before and during the first week of school. Teachers reported strong odors in the classrooms, and forty-one individuals, including students and teachers, reported adverse health effects in the first three days of school. Then another lindane application was made in several classrooms after school one day, and again just before students arrived the following morning. The school was closed later that day due to continuing health complaints and the lingering odor of the chemicals. A total of ninety-eight health complaints were received and reviewed by the Louisiana Office of Public Health (LOPH). Symptoms reported by children and adults included headaches, abdominal pain, diarrhea, nausea, skin rashes, difficulty breathing, and sore throats.

The school remained closed for weeks while three state agencies investigated the illegal applications. The presence of pesticides was confirmed by analysis of wipe samples from classrooms and the playground. The National Guard was called in to help with decontamination (cleaning of classrooms and removal and replacement of playground soil and sod). Portable classrooms that had been directly treated with lindane were torn down.

The LOPH concluded that children were exposed to pesticides by inhaling vapors when they entered treated classrooms, and possibly via hand-to-mouth contact and skin absorption from touching residues on desks and teaching materials. The agency also concluded that the health symptoms reported were precipitated by pesticide exposure. Ironically, the LOPH report about the incident also noted the "the flea infestation remained a problem in the school, even though copious amounts of pesticides had been used."

The parish school board was fined $2,500 for violating state pesticide laws. A class-action lawsuit filed by parents against the school district was settled in 1998. The district spent nearly a million dollars for soil testing, cleanup, and rebuilding.[49]

June 1993, Forestville, New York. Forestville Central High School was evacuated and closed for a day in late May following the application of a "weed-and-feed" product. Vegetation control containing 2,4-D was applied to lawns around the school. Odors were drawn into school via the ventilation system. The district was cited for allowing pesticides to be applied by an uncertified applicator, and was fined $500. Then, just a few months later, a custodian under the direction of the school nurse sprayed Rid Lice Control Spray (permethrin) in an elementary classroom on the same central school campus. The school district was again cited for allowing pesticide application by an uncertified applicator, and signed a consent order waiving a $1,200 penalty.[50]

September 27, 1993, Montgomery County, Pennsylvania. Seventeen children were sent home from Montgomery Elementary School just after lunch with headaches, nausea, vomiting, diarrhea, and low-grade fevers. Food poisoning was ruled out, as the children ate different things. One alert parent noted that her son's flu-like symptoms (headaches, stomachaches, and low-grade fevers) returned when he went back to school the following week. She also noted that his symptoms seemed to occur when he was in the cafeteria or after lunch, but cleared up over the weekend. Then she learned that at least one teacher and eleven other students were also experiencing frequent headaches, stomachaches, and low-grade fevers at the school. One girl suffering from the symptoms even had a grand mal seizure.

The parent began to investigate further after her son's doctor suggested she call state agencies about having the school tested for environmental contaminants. She learned that the school was making regular applications of Dursban (chlorpyrifos) in the kitchen, cafeteria, and teacher's lounge in an effort to control ants, and that the insecticide was applied at the school on September 27, the day so many children got sick. The parent reported that she was given different stories about the time of day that the application occurred, but she believed that it was during the school day.

The parent asked her son's doctor to do a blood cholinesterase test, and results indicated a recent exposure to organophosphate pesticides. The county health department took air samples at the school nineteen days after the pesticide application. Samples were collected in the cafeteria and in a classroom, where windows were opened for ventilation during the test. The test failed to find pesticide residues. (In fact, the tests that were done were not designed to detect organophosphate pesticides. Experts consulted about the testing say that it was conducted improperly, and furthermore, that any pesticide residues that remained at that point would likely have been absorbed into the carpet.) No other environmental tests were done, nor were human blood or urine samples collected. The agency concluded that there was no evidence of a public health hazard. No attempt was made to determine the cause of ongoing symptoms of illness reported by students or the teacher. According to a letter from his doctor, one boy remains chemically sensitive.

The state Department of Agriculture lost or inadvertently destroyed its files on this case. However, personal notes by one investigator said the agency concluded that the illnesses at the school on September 27 occurred before the pesticide application was made that day, and that no pesticide violations were found. The state health department did not get involved in the investigation.[51]

Fall 1993 to Spring 1994, Indiana. Eighth-grader Emily Schultz was diagnosed with non-Hodgkin's lymphoma in the fall of 1993. In struggling to find out what could have caused their young daughter to contract this deadly disease, her parents learned that studies have found that people exposed to 2,4-D and other phenoxy herbicides have been shown to have elevated rates of this cancer. Then they discovered, much to their horror, that their daughter's school district was routinely using this very herbicide to kill dandelions and keep its school grounds looking neatly groomed.

Emily's cancer was brought into remission by a grueling course of chemotherapy. However, on the girl's first day back at school in the spring, the school district made another application of herbicides to the school grounds. When Emily's mother arrived to pick up her daughter that afternoon, she was horrified to smell the chemical odor, and appalled to realize that she had brought Emily into contact with the chemical while she was in a weakened and vulnerable condition. Sadly, the girl's lymphoma returned within the month. She died before summer's end that year. Having failed to identify any other known risk factors relevant to their daughter's illness, Emily's family believes that exposure to 2,4-D-containing weed killers which were used at her school may well have caused or contributed to her initial illness, her relapse, and her eventual death.[52]

December 1992, Ashtabula County, Ohio. Maintenance staff at a school for multiply handicapped children used an old bottle of the insecticide malathion, spreading it around the perimeter of a small shed in an effort to control rodents. They applied the chemical on a Wednesday night after school was out. The next morning the insecticide vaporized, and winds carried the fumes into a room where students and parents had gathered for a holiday play. Many people noticed the odor, and several staff members complained of nausea and sore throats. By noon, complaints and "strange maladies" increased, including excessive salivation, tearing, nausea, fatigue, and headaches. Part of the school was evacuated, but vapors then entered other areas of the building via windows and heating intake ducts. At least two people went to private physicians because of health complaints associated with the exposure, and their physicians validated their conditions. The Ashtabula County Health Department later concluded that the symptoms experienced were most likely related to exposure to the malathion's petroleum distillate base.

The school was closed the next day while air testing and cleanup began. That Monday the shed and the contaminated soil around it were removed; they still carried an overpowering stench from the pesticide. Air samples taken in the school on Monday morning showed no traces of the insecticide, and school officials planned to reopen the school, but state health department officials suggested that workers wash every surface in the school three times to ensure that no traces of the chemical or its petroleum base remained. Ultimately, the school was closed for more than a week, and cleanup and waste removal cost more than $15,000.[53]

December 7, 1992, St. Paul, Minnesota. Four students and three adults from Woodbury High School were treated at a nearby emergency room after they were exposed to the insecticide malathion. Emergency room personnel examined an additional twenty-seven students. The students reported light-headedness and a teacher reported a headache. The incident occurred when a student mixing a spray for use on plants in the school's greenhouse spilled about half a cup of it. He used his bare hands to wipe up the spill. However, the solution evaporated and fumes quickly spread into an adjacent classroom and hallway. Students were immediately evacuated, and the fire department was called. The state health department does not have the file on this case.[54]

October 26, 1992, Eastchester, New York. Children, teachers, and other staff of Eastchester High School noticed a strong odor and experienced headaches, nausea, and eye and respiratory irritation immediately following their return to school on Monday morning. Some children developed rashes, sore throats, and other symptoms. The school had been sprayed over the weekend for roach control with the pesticides Empire 20 (chlorpyrifos), Vectrin (resmethrin), and diazinon. A boric acid paste was also applied. The applications were part of routine pest control used throughout the school district.

The school was closed later in the day. It was ventilated and reopened for part of the next day, but then closed again due to continuing strong odors. A professional cleaning firm was hired to conduct a massive cleanup, including new paint and floor caulking in some areas. However, air and surface wipe samples taken after the cleaning showed the chlorpyrifos was still present in many locations, so another cleaning was done. Even after the second cleaning, small amounts of pesticide remained, but a decision was made to reopen the school. Ultimately, the school was closed for almost three weeks as crews worked to clean up the pesticide residues. The pest control firm that made the application was cited for numerous violations, and their business license was revoked. A state and county health department report on the incident concluded that the symptoms seen among students, teachers, and staff were consistent with exposure to the pesticides. A newspaper account quoted a county health official as saying that inhalation and dermal exposure to the "inert" petroleum distillates in one of the products were the likely cause of most of the symptoms. Several lawsuits resulted from this incident. Parents noted that the roaches returned to the school even before the students did.[55]

January 21, 1992, Saddle Brook, New Jersey. Scores of children complained of sore throats, headaches, difficult breathing, nausea, vomiting, and rashes and were sent home in the days after End-Sect Insecticide (resmethrin) was applied by school maintenance workers in a crawl space underneath a first-grade classroom during school hours. Another chemical, End-Sect Vaporizer (pyrethrins and piperonyl butoxide), had been applied by a night custodian just a week earlier around the sink in the same classroom. Both applications had been made in an effort to control termites. School employees who were not licensed pesticide applicators applied the chemicals, which were no longer legally registered for use. The chemicals were both stored in thirty-gallon drums, one in a crawl space under the school, and one in a garage at another school. Of note, the area under the first-grade classroom had been treated for termites with fifty-five gallons of another insecticide by a commercial pest control firm just nine months earlier. Another part of the school was also treated with Orthene by a second pest control firm just the day after the January 21 application. Another application of 120 gallons of a termiticide was also made under the kindergarten room on February 1.

A parent filed a complaint with the state on Friday, January 24, and a state inspector arrived at the school to do an inspection and testing on Monday, January 27. After samples were collected, school officials elected to close the school pending results. A swab sample collected near the classroom sink showed residues of pyrethrins, which had been applied there eleven days earlier. Six samples from the classroom and crawl space were analyzed only for petroleum distillates, not for the active ingredients of the pesticide products used. The inspector noted that there were no established air standards for the active ingredients of the products used, pyrethrin or resmethrin. The sample from the crawl space tested positive for petroleum hydrocarbons, while the sample from the classroom did not show detectable levels of petroleum hydrocarbons six days after the crawl space application.

The school reopened on February 3, after a four-day closure while the chemical was cleaned. School board trustees were fined nearly $6,000 by the state Bureau of Pesticide Compliance for ordering pesticide applications to be made by unlicensed employees, and for illegal use of a cancelled pesticide product. One parent filed a notice of intent to sue in an effort to cover medical expenses related to surgery her six-year-old had to remove gum boils the parent says were related to the exposure. Other children apparently also developed gum boils the week of the incident. The state health department has no case file on this incident.[56]

May 8, 1991, Coral Springs, Florida. Thirty-four students and eight adults were sent to area hospitals and ten others were treated by paramedics at Forest Hill Elementary School the day after being overcome by strong pesticide fumes. Symptoms reported included churning stomachs, dizziness, and a bad pepper-like taste in the mouth. The school had been sprayed the night before with two synthetic pyrethroid insecticides, Tempo 20 WP (cyfluthrin) and Micro-Gen ULD BP-100 (pyrethrins and piperonyl butoxide). Investigators suspected that some of the insecticide had landed on top of steamers or ovens in the cafeteria, and later vaporized when the ovens were turned on, resulting in the sickening fumes. All 175 schools in the Broward County school district were sprayed regularly with these same chemicals in an ongoing effort to control roaches, ants, and fleas.[57]

May 5, 1989, Cross Lanes, West Virginia. Andrew Jackson Junior High School was closed after four years of complaints by teachers and students of persistent coughs, fatigue, headaches, respiratory problems, nausea, and numbness in their limbs. Federal investigators found the cancer-causing pesticide chlordane in the air at levels eleven times higher than the federal evacuation limit. The chemical was applied at the school to combat termites. The school district paid $600,000 in 1995 to settle a lawsuit brought by sixty-seven students and school employees who said they experienced nerve damage, immune system problems, bone marrow dysfunction, aching joints, allergic reactions, and cancer resulting from the exposure. The exterminator paid more than a million dollars in fines. The school was reopened in February of 1990 after an extensive cleanup.[58]

1989 to 1990, Greenville County, South Carolina. After a parent inquiry, state investigators found a pattern of illegal pesticide applications in Greenville County schools, including fogging of classrooms with the restricted-use pesticide lindane (in an effort to control head lice), indoor use of agricultural formulations of diazinon, and applications by non-certified school maintenance personnel.[59]

April 28, 1987, Grand Island, New York. The local fire department was called in to evacuate Kaegebein Elementary School when strong pesticide odors entered a classroom after plants in a solarium inside the school foyer were sprayed with an over-the-counter malathion product. The spraying was done by two volunteers at the school who were also members of a local garden club. They were attempting to control mealy bugs on the plants. The school was reopened the following day after a cleanup and air sampling by the health department. A warning letter was sent to the school for allowing the application to school property by unlicensed individuals.[60]

October 2, 1986, Honolulu (Oahu), Hawaii. At least thirty children and three adults at Waianae Elementary School complained of headaches, stomachaches, breathing difficulties, dizziness, nausea, and other symptoms. The insecticide Dursban 4-E (chlorpyrifos) was applied by the state health department around the perimeter of certain school buildings the afternoon before in an effort to control fleas present from dogs sleeping under the portable classrooms. Another application had been made just two weeks earlier. Health department investigators found "no evidence of pesticide misuse." However, the agency's epidemiologist stated in a letter that the evidence indicated that health symptoms may have been caused by solvents (xylene) and other ingredients (diethyl sulfides) in the pesticide. The school remained closed the following day. An inspection done after the second treatment found that fleas were still present. Following this incident, the school installed screens around crawl spaces to prevent access by dogs.[61]

January 28, 1987, Silver Creek, New York. A school custodian under direction of a school nurse sprayed Diatox C (diazinon) on carpeting in four classrooms at Silver Creek Elementary School in an attempt to control head lice. The over-the-counter product that was used was not registered for use in New York. Though the application was made on a Saturday, strong odors lingered when classes resumed on Monday. Despite cleaning efforts, the rugs eventually had to be removed. The classrooms were unusable for several days. The district was cited for applying a pesticide inconsistent with its label (it was not labeled for head lice control), and for allowing the application to be made by an unlicensed applicator, among other violations.[62]

April 1992, Tucson, Arizona. On April 24, 1992, more than 100 people in Tucson, Arizona, including firefighters, police officers, paramedics, nurses, and physicians, responded to reports that students at an elementary school had been exposed to an unknown and possibly toxic substance; 296 students were transported to eight hospital emergency departments. None were diagnosed as acutely ill. The substance was identified as approximately twenty-two milliliters of malathion diluted in fifteen liters of water and applied with a hand-held sprayer about 100 meters from the school. The odor apparently drifted to the school on winds of nineteen to twenty-four kilometers per hour. The episode was determined to be epidemic hysteria, possibly triggered by the malathion odor, but perpetuated by the stress of the emergency response.[63]

June 1997, Fontana, California. Janine Matelko's daughter Chrissy Garavito, age fifteen, died on June 30, 1997, allegedly due to ongoing exposures to organophosphate pesticides heavily sprayed throughout her school district in Fontana, California. Chrissy had been having seizure-like episodes while enrolled as a student at South Ridge Middle School. She would turn blue and stop breathing. Her physician placed her on an anti-seizure medication. Happily, these episodes abated when Chrissy transferred to Fontana High School, where she was freshman and sophomore class president, a cheerleader, a star athlete, and an honor-roll student. In late June, Chrissy returned to the pesticide-laden South Ridge Middle School to play in an all-star softball game. She and her teammates changed clothes in the locker room and then took the field, but Chrissy collapsed while sliding into home plate. It took paramedics twenty minutes to get a pulse. She was transported to a local hospital and kept on life support for about a week until she died on June 30, 1997.[64]

Herbicides on School Grounds

Consider these startling facts about three herbicides that are commonly used on school grounds and are widely believed to be "safe" and to break down rapidly into harmless components.

Glyphosate, the active ingredient of Roundup, has been called "extremely persistent under typical application conditions" by the EPA. Tests have shown that it can persist in soil for up to three years. Glyphosate has also been shown to cause genetic mutations in tests on human, animal, and plant cells.

The EPA has stated that chronic exposure to lawns treated with oryzalin (the active ingredient of Surflan) "is of concern because oryzalin is a carcinogen and persistent. There is a potential for continued, substantial contact with treated surfaces, especially among children. There are no data to evaluate potential exposure to turfgrass, and therefore the safety of this use cannot be evaluated."

Dichlobenil (the active ingredient of the herbicide Casoron) can persist in soil for up to five years. It kills weeds by continuously emitting a toxic vapor into and above treated soil. It also causes cancer in animals, and is classified by the EPA as a possible human carcinogen.[65]

Playground Toxins

Until recently, approximately 90 percent of outdoor wooden structures were made using wood treated with chromated copper arsenate (CCA). This substance, which is roughly 25 percent arsenic, prevents rot and repels pests that might damage wood.

In the 1980s, almost all industries were told by the EPA that they could no longer use arsenic in their consumer products. But the lumber industry caught a huge break from the federal government, and wood producers were granted an exemption from the new law. Because of this, picnic tables, playground equipment, and other wooden structures currently in place at schools and in park playgrounds may be constructed of wood that contains high levels of arsenic.

Arsenic can still seep out of wood, ending up on palms and fingers. This exposure is a concern for young people. Even after the toddler stage, children can transfer arsenic from wood to the food they eat, or otherwise inadvertently ingest it. Finally, children in general are more vulnerable to arsenic, which has been linked to various cancers, organ diseases, and neurological problems.

Tests performed in 2001 by the EPA revealed that lumber sold in major stores such as The Home Depot and Lowe's Home Improvement contained arsenic far in excess of the guidelines set by the EPA. That so-called safe level is ten micrograms of arsenic per liter of drinking water. On average, surface contamination of an area of CCA-treated lumber the size of a child's hand exceeded that level by 120 times.

Early 2002 marked a turning point, when the EPA and lumber industry representatives decided to phase out use of CCA-treated wood by 2004. While this was great news for parents and other concerned citizens, it is not the end of the story. For one thing, a phase-out does not do anything to remove the threat of wood still available in store inventories. Keep in mind that 90 percent of outdoor wooden structures used CCA-treated wood, which is a huge amount of wood that can still expose children to arsenic ingestion.[66]

Pesticides and Asthma

Asthma is a chronic, potentially fatal inflammatory disease of the respiratory system. Nearly one-third of people with asthma are children. Asthma is the number one cause of hospitalization and chronic health conditions among children, and is the leading cause of school absenteeism. Every year, asthma accounts for fourteen million lost days of school. In 2001, more than five million children aged five to seventeen in the United States were reported to have a current diagnosis of asthma.[67]

Asthma can be triggered by pesticides. Several types of pesticides are known to cause allergic reactions or airway constriction, including pyrethrins, pyrethroids, organophosphates, and carbamates.[68] Studies indicate that exposure to organophosphates disrupts the part of the nervous system that regulates the motor functioning of the lungs. This had lead researchers to hypothesize that pesticides are among the preventable causes of asthma in children. Unfortunately, pesticide use in schools is widespread. Four of the five pesticides most commonly used in California schools, cyfluthrin, diazonin, glyphosate, and pyrethrins, have been linked to asthma and other respiratory problems. Cyfluthrin can cause irritation of the nose, throat, and upper respiratory tract, leading manufacturer Bayer Corporation to state, "Persons with a history of asthma, emphysema, or hyperactive airways disease may be more susceptible to exposure."[69] Diazonin causes acute symptoms, including wheezing, coughing, and pulmonary edema (fluid in the lungs).[70] Glyphosate can cause the accumulation of excess fluid in the lungs. Studies show that glyphosate can persist in soils for up to a year. Pyrethrins contain allergens that cross-react with ragweed and other pollens. People with asthma can have severe reactions to pyrethrins.[71]

A bill introduced in the California assembly would have banned the use of the most highly toxic pesticides in schools, including many that have been linked to asthma, such as pyrethroids, carbamates, and organophosphates. However, the bill died in the 2003–2004 session and has not been reintroduced.

Legal Aspects of Asthma at School

Students suffering from asthma triggered by pesticides or uncontrolled pest populations may be able to use the Americans With Disabilities Act (ADA) of 1990 to require schools to provide non-toxic, effective pest management. Fortunately, the use of the ADA as a remedy for environmental disabilities does not conflict with FIFRA. Furthermore, reduction in asthma attacks would increase school attendance rates and prevent what is essentially the denial of a vital public service—education.

The rights of students with disabilities are defined under three federal laws: the Individuals Education Act (IDEA), Section 504 of the Rehabilitation Act of 1973, and the ADA. These rights are also covered under certain state statutes and regulations. Federal rulings on specific cases continue to clarify what these laws mean for students with asthma. A child does not have to be classified as "special needs" to qualify for accommodation or special planning, such as an individualized health plan. The ADA incorporates and extends the rights and responsibilities of Section 504 of the Rehabilitation Act of 1973 to include public services and places of public accommodation, such as preschools, daycare centers, and private schools.

Asthma is a condition that is considered a disability under the ADA. A disability is an impairment which affects a person's respiratory system and which substantially limits one or more of his or her major life activities. If a person has a record of that impairment, he or she is regarded as having that impairment. Asthma is a physical impairment that impairs what Department of Justice regulations consider to be a "major life activity," including breathing.[72]

In *Alvarez v. Fountainhead, Inc.,* the plaintiff was a student at a private Montessori school who used Title III of the ADA to obtain a reasonable accommodation to control his asthma.[73] Title III prohibits a place of public accommodation from discriminating against an individual on the basis of a disability (public schools fall under Title II) and entitles a disabled person to the protections of Title III of the ADA.[74] Under Title III, similar to Title II, "disability means, with respect to an individual, a physical or mental impairment that substantially limits one or more of the major life activities of such individual."[75] The court found that the plaintiff, who suffered from asthma, was a "person with a disability" and was entitled to the protections of Title III.[76] Discrimination is defined under Title III to include a denial of the opportunity to participate in, or benefit from, a public accommodation's goods and services.[77] Title III requires a public accommodation to make "reasonable modifications in policies, practices, or procedures" when such modifications are necessary "to ensure full and equal enjoyment of its services by individual with disabilities."[78]

Danielson, Connecticut

A student and his family requested that his school stop using harmful pesticides on the high school's football field and surrounding grounds, which caused the student to become ill, resulting in his inability to attend school for a total of eight weeks. The student was then placed on a Section 504 Plan, which gave the student the option of staying home when he felt ill from chemical exposure and provided tutoring. However, the student and his family complained that the tutoring was inadequate and that the student missed out on other educational opportunities and social events at the school. The student's family filed a complaint under Section 504 of the Rehabilitation Act of 1973 and Title II of the ADA. (Section 504 prohibits discrimination on the basis of disability by a recipient of federal funds from the U.S. Department of Education, and the ADA prohibits discrimination on the basis of disability by public entities.) The U.S. Department of Education, Region I, stated, "The ADA has essentially extended the anti-discrimination prohibition embodied under Section 504 to all state and local governmental entities, including public school systems."[79] In a letter to resolve the complaint, the Danielson, Connecticut, school district attached an addendum to the student's Section 504 Plan, stating that the school district would use alternatives to toxic pesticides that caused the student to become ill. This situation is an example of an administrative remedy that cited the ADA as the enabling authority to offer relief to a student whose illness was caused by school pesticide application. Unlike *Alvarez*, no court action was required to provide relief under ADA.

A student learns by using his or her central nervous system, assisted by a healthy body, adequate nutrition, a positive sense of well-being, good teachers (both inside and outside of school), and a clean environment. Pesticide exposure, however, robs a student of a clean environment, can undermine or destroy the student's health, and may directly affect the student's central nervous system. Learning then becomes another casualty of pesticide exposure.[80]

The Pesticide Industry's Position—Then and Now

With the publication of *Silent Spring* in 1962, Rachel Louise Carson set off a nationally publicized struggle between the proponents and opponents of the widespread use of poisonous chemicals to kill pests. Miss Carson, an opponent, was subjected to a torrent of criticism. Here are a few examples:

Dr. Robert White-Stevens, a spokesman for the pesticide industry, said, "The major claims of Miss Rachel Carson's book, *Silent Spring*, are gross distortions of the actual facts, completely unsupported by scientific, experimental evidence, and general practical experience in the field. Her suggestion that pesticides are in fact biocides destroying all life is obviously absurd in the light of the fact that without selective biologicals, these compounds would be completely useless. The real threat, then, to the survival of man is not chemical but biological, in the shape of hordes of insects that can denude our forests, sweep over our croplands, ravage our food supply, and leave in their wake a trail of destitution and hunger, conveying to an undernourished population the major diseases, scourges of mankind."[81]

An individual from California wrote to the *New Yorker*, which originally serialized her work in several articles: "Miss Rachel Carson's reference to the selfishness of insecticide manufacturers probably reflects her Communist sympathies, like a lot of our writers these days. We can live without birds and animals, but, as the current market slump shows, we cannot live without business. As for insects, isn't it just like a woman to be scared to death of a few little bugs! As long as we have the H-bomb everything will be O.K. P.S. She's probably a peace-nut, too."[82]

Forty years later, in a somewhat more sophisticated vein, the executive director of Responsible Industry for a Sound Environment (RISE) sent the following form letter to thousands of school districts in the United States, urging them to use pesticides:

"Dear [School Administrator],

"If it's your responsibility to maintain a clean, safe, and healthy school environment, you know the important role pesticides play as part of your overall facility management. Leading researchers, scientists, and even doctors like C. Everett Koop [no mention is made of the fact that the former surgeon general was strongly supported by chemical companies] agree that pesticides pose no risk to the health of children or adults when used according to label instructions.

"I'm sure that you will agree that pesticides are a valuable tool in protecting the health and safety of children on your school properties. At times, you may find it important to let others know pesticides protect children from the health and safety risks posed by pests such as cockroaches, rodents, poison ivy, and lice. If or when a pesticide issue arises in your school, we invite you to share these facts with parents, teachers, and students:

1. **Without pesticides, pests pose a serious health and safety risk to children and adults.** Cockroaches, ants, flies, fleas, lice, mosquitoes, ticks, wasps, and rodents are serious health concerns due to their bites, stings, and ability to transmit diseases.

2. **Pesticides are extensively tested and highly regulated.** Every pesticide must successfully complete as many as 120 government-mandated tests before the Environmental Protection Agency considers label approval and product registration. The entire development and testing process takes eight to ten years at a manufacturer's cost of $35 to $50 million or more per product. On average, only one in 20,000 potential products ever make it to the marketplace.

3. **Integrated Pest Management (IPM) is best.** To control pests, methods such as sanitation, structural repair and maintenance, watering, mowing practices, and judicious use of pesticides should be used. A balanced approach, which included the use of pesticides such as necessary, is one that will assure health and safety for children and adults. Pesticides should not be considered for

emergency use only. The purpose of responsible pest control is to prevent emergencies, if your school has implemented IPM, promote it!"[83]

The following response shows what New York Attorney General Eliot Spitzer (now governor-elect) and the state's Department of Education thought of the above letter:

"Attorney General Eliot Spitzer has asked the Department to share with school administrators the following information. In September 2000, a pesticide industry group known as RISE wrote to more than 20,000 school facilities nationwide, including those in New York state. The materials distributed by RISE promote pesticide use with deceptive claims and irrelevant anecdotes about the health and environmental impacts of pesticides. The Attorney General wants you to be fully informed of the correct information so that you can make pest control decisions to best protect the health of students, staff, and visitors.

"While ostensibly promoting Integrated Pest Management (IPM) at schools, the materials sent by RISE actually encourage continued excessive reliance on pesticides by schools. This central role for pesticides runs contrary to the positions of the Department, the Attorney General's office and countless other governmental and citizens' groups. However, it is understandable that RISE advocates this role, given that its mission, as set forth at its Web site, is to: 1) provide a strong unified voice for the specialty pesticide industry; 2) positively influence public opinion and policy; and 3) promote the use of industry products.

"In advocating pesticide use, RISE makes numerous deceptive safety or irrelevant claims. For instance, that claim that 'pesticides pose no risk to the health of children or adults when used according to label instructions' is not only false, but is specifically prohibited by federal regulations from appearing on the label of any pesticide product. The reference to the West Nile virus in New York City is largely irrelevant to school settings, given that the virus victims were elderly, and transmission is believed to have occurred in the evening during the summer. Similarly, the malaria reference is irrelevant given no infected mosquitoes were ever found.

"The Attorney General is concerned that RISE's mailing will be relied upon by school facilities managers and administrators, and its deceptive claims might be repeated to parents, students, and school staff. . . .

"While the public should not simply accept the risks associated with severe infestations at schools, it is not necessary to expose our children to highly toxic substances in the name of pest control. Properly planned and implemented IPM programs can serve to control pests without introducing toxic materials into the school environment."[84]

The Frontline Interview

In February 1998, Douglas Hamilton, producer of *Frontline*'s "Fooling with Nature," interviewed Dawn Forsythe, former manager of government affairs for Sandoz Agro, Inc. (now Novartis AG), a pesticide manufacturer based in Basel, Switzerland. Ms. Forsythe was the company's sole lobbyist for the entire United States and headed the pesticide industry's first committee on endocrine disruption before she left Sandoz at the end of 1996. Here are her candid remarks:

"It is their science or it's no science. They are so intricately involved with pesticides that who is anybody to tell them that pesticides can react in a way that is totally unsuspected? It's hard to describe the personal attachment that many in the industry have towards the industry. They grow up as a pesticide salesman or a bench scientist and they climb up the ladder. You don't move from the pesticide industry to another industry usually. I know that our CEO started out as a salesman. It is something that is their life, and when an issue comes up that tries to show them that their whole life may be a lie—I would have problems with that. You have to justify it to yourself. You have to believe that you are not intentionally putting children or women or men in danger. And they are not intentionally doing it. But the road to hell is paved with good intentions."[85]

More from Industry

Representatives of the $1.5 billion non-agricultural pesticide industry, which makes the herbicides and pesticides used in school applications, say their critics overstate the potential harm of their products to children and underestimate the public health threat of the bugs and weeds they are designed to kill.

"Pesticides are the most efficient and effective means of protecting children's health in schools," said Allen James, president of RISE. James referred to a 1997 National Institute of Allergy and Infectious Diseases study that included cockroach allergens among the top causes of childhood asthma. It also mentioned food infestations, poison ivy, and weed-strewn playgrounds as prevalent threats to students' health and safety.[86]

In June 2003, the giant Grocery Manufacturers Association, the world's largest association of food, beverage, and consumer product companies, presented the industry's position in a letter from Director of State Affairs Kristin Power to California State Senator Mike Machado, opposing a bill regulating pesticides then in the legislature:

"Dear Senator Machado:

"On behalf of the Grocery Manufacturers of America, I am writing to express our opposition to Assembly Bill 1006 . . . scheduled for hearing on Tuesday, July 1.

"GMA and its member companies have worked collaboratively with members of the California legislature over the past several years to address concerns regarding the use of pesticides on school campuses. We appreciate the author's [Senator Chu, the bill's sponsor] concern about the issue, however AB 1006 does not take into consideration the risk of pests and allergens and the benefits of pesticides and cleansers. Children are particularly vulnerable to bites and stings from spiders, ticks, wasps, and fire ants. Additionally, many suffer allergic and asthmatic reactions to the presence of roaches, bacteria, and other contaminants found in schools.

"AB 106 would prohibit the use of many products that are safety and effectively used to disinfect and prevent insect infestations, including mold and mildew removers, cleansers, and insecticides. These products are appropriately labeled for safe use and are strictly regulated under both federal and state laws."[87]

Ralph Engel, president of the Chemical Specialties Manufacturers Association, said parents should be notified about school pesticide use, but he opposed any suggestions that pesticides use should be reduced in schools. RISE's Allen James said the proposed seventy-two-hour notification would interfere in "the timely use of pesticides."[88] The American Public Health Association quickly responded:

"There are more than 54 million children and 2.3 million teachers in our K-12 schools today. Every child and every teacher should have a right to go to a school that is environmentally clean, safe, and well designed. They have a right to a school learning environment with fresh clean indoor air, safe water, exposure to the out-of-doors and exercise and where health risks from toxicides such as chemical pesticides would be a non-factor."[89]

A Few Caveats

One major concern is that most school employees are not familiar with pesticides, proper usage, and potential problems. Some states require trained personnel to apply pesticides, but other states have no oversight.

Pesticides exposure is also known to affect the cognitive and motor skills of students. The U.S. Office of Technology Assessment reports:

"In general, [human health] research demonstrates that pesticide poisoning can lead to poor performance on tests involving intellectual functioning, academic skills, abstraction, flexibility of thought, and motor skills; memory disturbances and inability to focus attention; deficits in intelligence, reaction time, and manual dexterity; and reduced perceptual speed."[90]

A Bill of Rights

The following Bill of Rights was adopted by the New York State Board of Regents in June of 1994:

"Every child and school employee has the right to an environmentally safe and healthy school which is clean and in good repair.

"Every child, parent, and school employee has a right to know about environmental hazards in school environment.

"Schools should serve as role models for environmentally responsible behavior.

"School officials and appropriate public agencies should be held accountable for providing an environmentally safe and healthy school facility.

"Federal, state, local, and private sector entities should work together to ensure that resources are used effectively and efficiently to address environmental health and safety concerns."[91]

A Teacher's Warning

Irene Wilkenfeld, a former teacher who became chronically ill after being exposed to pesticides decades ago, issued a final, poignant admonition:

"We are sabotaging our children's success in school with our ignorance, our inertia, and our silence. Until school-based environmental exposures are substantially curtailed, our nation's youngsters will continue to fall short of our educational goals. Given the barrage of sobering statistics, why would intelligent and caring people intentionally and regularly saturate their schools, homes, workplaces, and lawns with these toxic chemicals? It makes no sense to use poisons that impair a child's ability to think and develop normally in the very places that are mandated to provide a safe learning and growing environment. We've been bingeing on pesticides for too long. It's time to get off the toxic treadmill."[92]

Irene Wilkenfeld died of liver failure due to an exposure to the hepatotoxic termiticide chlordane. She was exposed at a school where she taught in the 1960s, and spent most of the rest of her life teaching people about the dangers of pesticides and the importance of safeguarding the school environment.

No Need for Pesticides

When an IPM approach is part of school pest management, pesticides may be used as a last resort. However, principals could attempt another initiative, which would be called the "Prevention and Cleanliness Plan for Pest Management." It would be composed of two steps. First, custodial staff would periodically monitor floors, school walls, and baseboards for sites that might harbor rodents and insects. Upon locating cracks or crevices, personnel would seal them with caulk or other appropriate sealants. Custodians would also double-check the cafeteria for food scraps at the end of the school day. Second, the principal would unveil a schoolwide No Food and No Drinks policy, including signs to that effect in all classroom, halls, and restrooms. The cafeteria would be the only place where food and liquid refreshment could be ingested. For successful implementation, this policy would require the full cooperation of students, teachers, and custodians. There is no reason why this plan would not be workable and could not be successful without having to utilize any harmful pesticides.

Notes

1. Susan Johnson, Deputy Agricultural Commissioner in Ventura County, CA, PBS's *NOW*, August 19, 2005.

2. "Pesticide Pact Aimed at School, Kids," *MSNBC News*, June 21, 2002.

3. *Integrated Pest Management for Schools* (Washington, D.C.: Environmental Protection Agency, June 2004): 1–5.

4. C. S. Miller, "The Compelling Anomaly of Chemical Intolerances," *Annals of the New York Academy of Science* 933 (March 2001): 1–23.

5. S. S. Addiss, N. O. Alderman, D. R. Brown, C. N. Each, and J. Wargo, "Pest Control Practices in Connecticut Public Schools" (North Haven, CT: Environment and Human Health, Inc., 1999).

6. J. Kaplan, S. Marquardt, and W. Barber, *Failing Health: Pesticide Use in California Schools* (San Francisco: CALPIRG Charitable Trust, 1998).

7. D. I. Volberg, M. H. Surgan, S. Jaffe, D. Hamer, and J. A. Sevinsky, *Pesticides in Schools: Reducing the Risks* (New York, NY: New York Office of the Attorney General, 1993).

8. *Human Health Issues* (Washington, D.C.: Environmental Protection Agency, n.d.).

9. "Keep Those Pests Away from School," *Alternative Medicine* (March 2002).

10. Brian Dementi, "Cholinesterase Literature Review and Comment," paper presented to EPA Scientific Advisory Panel Meeting of June 3–4, 1997, Arlington, Virginia, EPA Office of Pesticide Programs.

11. Ibid.

12. Ted Schettler, Jill Stein, Fay Reich, Maria Valenti, and David Wallinga, "In Harm's Way: Toxic Threats to Child Development" (Cambridge, MA: Greater Boston Physicians for Social Responsibility, May 2000).

13. Walter A. Alarcon et al., "Acute Illnesses Associated with Pesticide Exposure at Schools," *Journal of the American Medical Association* 294 (July 2005): 455–465.

14. "Pest Management Is People Management," *Pest Press* 9 (November 2004).

15. *Integrated Pest Management (IPM) in Schools* (Washington, D.C.: Environmental Protection Agency, n.d.)

16. *Step 2 in Establishing an IPM Program in Schools—Educating IPM Participants* (Washington, D.C.: Environmental Protection Agency, n.d.).

17. Gregory M. Williams, H. Michael Linker, Michael G. Waldvogel, Ross M. Leidy, and Coby Schal, "Comparison of Conventional and Integrated Pest Management Programs in Public Schools," *Journal of Economic Entomology* 98 (4) (2005): 1,275–1,283.

18. *Pest Control in the School Environment: Adopting Integrated Pest Management* (Washington, D.C.: Environmental Protection Agency, Office of Pesticide Programs, 1993).

19. J. L. Morehouse, C. W. Scherer, and P. G. Koehler, *Pests and Pesticides in Schools* (Gainesville, FL: University of Florida, March 1998): 2–3.

20. EPA. Op. cit.

21. W. Forbes, "From Spray Tanks to Caulk Guns: Successful School IPM in Montgomery County, MD," *Journal of Pesticide Reform* 10 (4) (1991): 9–11.

22. *Cost of IPM in Schools*, fact sheet from the Safer Pest Control Project (Chicago, IL: Safer Pest Control Project, 1998).

23. Marc Lame, IPM specialist, Monroe County school district, Indiana, personal communication, September 1998.

24. Angelo Ranieri, building engineer, Susquehanna school district, New York, personal communication, October 1998.

25. *Clean Schools, Safe Kids: Striving for Safer Pest Control in North Carolina Public Schools* (Raleigh, NC: Agricultural Resources Center, Pesticide Education Project, 2003).

26. *Pesticides—Evaluating the Cost of IPM in Schools* (Washington, D.C.: Environmental Protection Agency, May 2003).

27. Kagan Owens and Jay Feldman, *The Schooling of State Pesticide Laws—2002 Update* (Washington, D.C.: Beyond Pesticides/National Coalition Against the Misuse of Pesticides, 2003).

28. "Progress in State and Local School IPM Programs," paper presented April 9 at the Fourth National IPM Symposium, April 8–10, 2003, Indianapolis, IN.

29. "Naturally Rockford: Illinois Parents Urge Schools to Start Year without Toxic Chemicals," *Rock River Times* (Rockford, IL), March 31, 2006.

30. MASPIRG Reports, *Protecting Children from Toxic Pesticides: A Guide for Schools, Day Care Centers and School Age Child Care Programs in Massachusetts* (Boston, MA: Massachusetts Public Interest Research Group, n.d.).

31. Michael C. Levenson, "Most Schools Not Complying with Pesticide Regulations," Associated Press, May 26, 2004.

32. "Learning Curve: Charting Progress on Pesticide Use and the Healthy Schools Act," *Environmental Health Reports*, May 2, 2002.

33. Fran Henry, "A Chemical Reaction for Local Schools," *The Cleveland Plain Dealer*, February 7, 2005.

34. *Healthy Schools: Getting Hazardous Pesticides Out of Our Schools* (Seattle, WA: A Washington Toxics Fact Sheet, March 2003).

35. *New York State School Pesticide Law* (Washington, D.C.: Beyond Pesticides, n.d.)

36. *Integrated Pest Management Notification Letter For Parents or Guardians* (McMurray, Washington County, PA: Peters Township School District, n.d.)

37. Kathryn M. Vail, *Suggested Guidelines for Managing Pests in Tennessee's Schools: Adopting Integrated Pest Management* (Knoxville, TN: University of Tennessee Agricultural Extension Service, July 8, 1997).

38. *The ATSDR Environmental Health Nursing Initiative* (Atlanta, GA: Department of Health and Human Services, Agency for Toxic Substances and Disease Registry, June 5, 2003).

39. C. Bearer, "Environmental Health Hazards: How Children Are Different from Adults," *Future of Children* 5 (2) (Summer/Fall 1995): 11–26.

40. H. L. Needleman and P. G. Landrigan, *Raising Children Toxic Free: How to Keep Your Child Safe from Lead, Asbestos, Pesticides and Other Environmental Hazards* (New York, NY: Farrar, Straus and Giroux, 1994).

41. *Safety and Health Topics: Hazard Communication* (Washington, D.C.: U.S. Department of Labor, Occupational Safety, and Health Administration, 2004).

42. "School Pesticide Use Not Top Concern for Parents, Study Says," *Pest Control Technology*, July 16, 2001.

43. "Youths Exposed to Pest Spray at School, Report Will Check Chemical Levels," *New Orleans Times-Picayune*, January 23, 1999.

44. Case # R 5-2312-99-OZ (Albany, NY: New York Department of Environmental Conservation, January 26, 2000).

45. "Health Agents to Discuss School Pesticide Spell," *Charleston Post and Courier* (South Carolina), March 17, 1999.

46. *List of Chemicals Evaluated for Carcinogenic Potential* (Washington, D.C.: Environmental Protection Agency, December 31, 1994).

47. "Chardon High Evacuates Classes to Escape Fumes," *The Cleveland Plain Dealer*, June 10, 1997.

48. "Ant Spray Use Forces Fort River Class to Move," *Daily Hampshire Gazette* (Amherst, MA), May 21, 1997.

49. "Second Pierre Part School Closed after Pesticide Contamination," *The Advocate* (Baton Rouge, LA), August 30, 1994.

50. Consent Order # R 9-4025-93-09 (Albany, NY: Department of Environmental Conservation, n.d.).

51. "Law Targets School Pesticide Use," *The Morning Call* (Allentown, PA), January 26, 1994.

52. Kathy Schultz, personal communication, December 1999.

53. Ray Saporito, " A Toxic Nightmare," *American School Board Journal*, September 1993.

54. "Woodbury Students Exposed to Pesticide," *St. Paul Pioneer Press*, December 8, 1992.

55. "School Weighs Risk of Pesticide," *The New York Times*, January 10, 1993.

56. "Pupils Moved Pending Tests for Pesticides," *The Record* (Hackensack, NJ), January 29, 1992.

57. "Insecticide Fumes Sicken Forty-Two at School," *The Miami Herald*, May 8, 1991.

58. "Kanawha School Board to Shell Out $600,000 to Settle Suit Over Pesticides," *The Charleston Gazette* (Charleston, WV), June 24, 1995.

59. "School District Broke Pesticide Regulations," *Greenville News* (Greenville, SC), May 12, 1990.

60. John Wainwright, personal communication, New York Department of Environmental Conservation, January 14, 2000.

61. "Flea Treatment Fumes Cause Early Dismissal at Waianae," *The Honolulu Advertiser*, October 3, 1986.

62. Consent Order # R 9-2040-87-03 (Albany, NY: New York Department of Environmental Conservation).

63. P. Baker and D. Selvey, "Malathion-Induced Epidemic Hysteria in an Elementary School," *Veterinary and Human Toxicology* 34 (2) (April 1992): 156–160.

64. Troy Anderson, "The Sick School Syndrome," *The Daily Bulletin* (Ontario, CA), June 1997.

65. *Intentional Poisons in Our Schools—Pesticides Pollute Classrooms and School Grounds* (Eugene, OR: Northwest Coalition for Alternatives to Pesticides, n.d.).

66. "Arsenic in Wood = Poison in the Playground," *Generation Green*, n.d.

67. L. Y. Wang, Y. Zhong and L. Wheeler, "Direct and Indirect Costs of Asthma in School-Age Children," *Prevention of Chronic Disease* (January 2005).

68. Cheng Shim and M. Henry Williams Jr., "Effects of Odors on Asthma," *The American Journal of Medicine* 80 (January 1986): 18–22.

69. Bayer Corporation, *Cyfluthrin Material Data Sheet*, 2002.

70. *Recognition and Management of Pesticide Poisonings*, 5th ed (Washington, D.C.: Environmental Protection Agency, 1999): 34, 38.

71. Caroline Cox, "Glyphosate Part 2: Human Exposure and Ecological Effects," *Journal of Pesticide Reform* (Winter 1995).

72. *Hunt v. St. Peter School*, 963 F. Supp. 850 (W.D. Mo. 1997).

73. *Alvarez v. Fountainhead, Inc.*, 55 F. Supp. at 850.

74. 42 U.S.C. Sec. 12181(7)(k).

75. 28 C.F.R. Sec 136.104 (2002).

76. *Alvarez*, 55 F. Supp. 2d at 1051.

77. 42 U.S.C. Sec 12182 (b)(1)(A)(i).

78. 42 U.S.C. Sec. 12182 (b)(2)(A)(ii).

79. *School Pesticide Incidents from Around the Country—A Selection of Student and School Staff Poisonings* (Washington, D.C.: Beyond Pesticides, n.d.): l.

80. U.S. Congress, *Neurotoxicity: Identifying and Controlling Poisons of the Nervous System* (Washington, D.C.: U.S. Government Printing Office, 1990).

81. John M. Lee, "'Silent Spring' is Now Noisy Summer," *The New York Times*, July 22, 1962.

82. Laura Orlando, "Industry Attacks on Dissent: From Rachel Carson to Oprah," *Dollars and Sense* 240 (March–April 2002).

83. Letter from Responsible Industry for a Sound Environment (RISE) to school administrators, August 31, 2000.

84. "Pesticide Industry Group RISE Communication," *School Executive's Bulletin* (Albany, NY: The New York State Education Department, Office of Elementary, Middle, Secondary, and Continuing Education, January 2001).

85. Dawn Forsythe, interviewed by Douglas Hamilton, producer of PBS's *Frontline*, February 1998.

86. John Nagy, "School Pesticide Question Challenges Policymakers," Stateline.org, October 13, 2000.

87. Kristin Power, letter to State Senator Mike Machado, June 12, 2003.

88. H. Josef Hebert, "Student Exposure to Pesticides Eyed," Associated Press, October 14, 2000.

89. "Children's Health Initiative," *APHA Policy Statement No. 20010* (Washington, D.C.: American Public Health Association, November 2000).

90. U.S. Congress, *Neurotoxicity: Identifying and Controlling Poisons of the Nervous System* (Washington, D.C.: Office of Technology Assessment, 1990).

91. *Bill of Rights: Environmental Quality in Our Schools* (Albany, NY: New York State Board of Regents, June 1994).

92. Irene Wilkenfeld, open letter: Pesticides Sabotage the Health, Behavior, and Academic Performance of Children, n.d.

Five

Pesticides in Homes, Lawns, and Gardens

There are substances commonly used in the home that make our lives easier. We use these substances in good faith, seldom questioning the fact that they could cause peripheral nerve or brain damage. Consumers rely on the government's and industries' judgment on health dangers associated with the use of chemicals and pesticides.

—Harold L. Volkmer[1]

Introduction

While we often consider our homes as sanctuaries—places of peace and safety—we may actually be living in danger zones filled with toxic airborne chemicals. Many of the building materials and housekeeping substances we use in our homes are air pollutants, capable of causing acute and long-term damage to our health, as well as the health of our pets. In fact, our animals are even more vulnerable than we are to the damaging effects of indoor air quality.

Indoor air pollution poses high risks to humans, especially sensitive groups, and has ranked among the top four environmental risks. Indoor air in homes is, on average, two to twenty times more polluted than the outdoor environment. Today, we are seeing new causes and mutations of disease as a result of the rapidly expanding development of the synthetic chemical industry. A staggering 900 chemicals are present in the average home environment. This soup of synthetic chemicals can affect multiple body systems and contribute to health-damaging effects to the upper respiratory tract, nose and sinuses, immune system, digestive system, reproductive system, central nervous system, and internal organs.

According to a recent survey, 75 percent of U.S. households used at least one pesticide product indoors during the past year. Products used most often are insecticides and disinfectants. Another study suggested that 80 percent of most people's exposure to pesticides occurs indoors, and that measurable levels of up to a dozen pesticides have been found in the air inside homes. The amount of pesticides found in homes appears to be greater than can be explained by recent pesticide use in those households.

Pesticides used in and around the home include products to control insects (insecticides), termites (termiticides), rodents (rodenticides), fungi (fungicides), and microbes (disinfectants). In 1990, the American Association of Poison Control Centers reported that some 79,000 children were involved in common household pesticide poisonings or exposures. In households with children under five years old, almost one-half stored at least one pesticide product within reach of children. Exposure to high levels of cyclodiene pesticides, commonly associated with misapplication, has produced various symptoms, including headaches, dizziness, muscle twitching, weakness, tingling sensations, and nausea. In addition, the EPA is concerned that cyclodienes might cause long-term damage to the liver and the central nervous system, as well as an increased risk of cancer.[2]

It is difficult to ignore the statistics: homeowners use 2 billion pounds of insecticides annually, both inside and outside their homes. In 2002, 3.2 million people reported medically related side effects from pesticides.[3] Each year poisonings result in nearly 900,000 visits to emergency rooms and some 1,100 deaths. The overwhelming majority of poisonings occur in homes. Many common household products can be poisonous, including pesticides, which can be dangerous if used incorrectly or if they are not stored properly and out of the reach of children. In 2003, children under the age of six were exposed to pesticides 50,415 times. However, experts estimate that this represents only one-fourth to one-third of pesticide exposure incidents reported to health-care professionals. A further challenge to collecting reliable pesticide exposure information is the fact that pesticide exposure may be misdiagnosed with symptoms of the common flu.

A survey by the EPA regarding pesticides used in and around the home revealed that almost half (47 percent) of all households with children under the age of five had at least one pesticide stored in an unlocked cabinet less than four feet off the ground, or within the reach of children. Furthermore, approximately 75 percent of households without children under the age of five also stored pesticides in an unlocked cabinet less than four feet off the ground. This number is especially significant because 13 percent of all pesticide poisoning incidents occur in homes other than the child's home.[4]

Indoor Air and Surfaces

It is becoming more widely recognized that most of our exposure to pesticides is through breathing indoor air and not through residues in our food. The EPA conducted a three-year study, 1986 to 1988, to estimate levels of exposure to selected household pesticides experienced by the general population.[5] Thirty-two different pesticides and breakdown products were detected at least once in air samples taken inside and outside the homes studied. The most frequently detected pesticides were the widely used household insecticides, chlorpyrifos, diazinon, and propoxur; orthophenylphenol, an active ingredient in disinfectants; and the now banned insecticide chlordane. Indoor air was found to have much higher concentrations of pesticides than outdoor air, a significant finding given that small children spend close to 90 percent of their time indoors. Overall, the study estimated that 85 percent of the total daily exposure to airborne pesticides was from breathing air inside the home.[6]

A study published in the *American Journal of Public Health* examined air and surface residues following indoor treatment for fleas with chlorpyrifos under the trade name Dursban. Three to seven hours after application, insecticide concentrations were found to be much higher in the infant breathing zone nearest the floor than in the more ventilated adult breathing zone. In addition, insecticide residues were found on the carpet twenty-four hours after application. Researchers estimated that the total amount of insecticide that infants would absorb, primarily through the skin, up to twenty-four hours after applications was ten to fifty times higher than what the EPA considers an acceptable exposure for adults.[7]

A review of thirty-seven children poisoned by organophosphate and carbamate pesticides in Dallas revealed that each child was exposed at home and nearly 70 percent of the cases occurred when a child ingested or drank improperly stored products.[8] In 15 percent of the cases, however, children developed symptoms thirty-six hours after the house was sprayed or fogged. The authors concluded that children's skin absorbs pesticides from contaminated carpets and linens.

In a pilot study of nine homes occupied by families with children between the age of six months to five years, pesticides were detected in all homes, with a total of twenty-three different pesticides detected in the study.[9] The number of pesticides found at each home ranged from eight to eighteen. The most frequently detected pesticides were chlordane, chlorpyrifos, dieldrin, heptachlor, and pentachlorophenol. The greatest number of pesticides and highest concentrations were found in carpet dust resulting from indoor treatment and track-in, potentially exposing infants and toddlers through dermal contact and oral ingestion.

Household Dust and Soil/Drift

At home or in daycare, small children spend considerable time on the floor, where they come in contact with and ingest dust and soil. Through normal play and hand-to-mouth activity, toddlers under the age of five ingest two and a half times more soil around the home than adults.[10] Overall, children were estimated to consume 0.01 grams to 1.3 grams of soil every day.[11]

Pesticides used around the home persist in dust, and those used on lawns, gardens, and nearby farms end up in soil and are tracked into the house on shoes and pets. Pesticides in soil and dust in indoor environments persist longer than they do outside, where exposure to sun and rain helps break down pesticide residues. In general, pesticides concentrate at higher levels in household dust than in soil.[12] One study measured the transport of lawn-applied herbicides to indoor carpet surfaces and carpet dust. Routine foot traffic across treated lawns brought herbicide residues into residences. Dirt tracked into homes via shoes transferred herbicides to carpet surfaces and carpet dust. Researchers estimated that 2,4-D would persist in carpet dust up to one year after lawn application.[13]

Children who live and play in agricultural areas are at higher risk of exposure to pesticides in dust and soil. Researchers in Washington state found that pesticide residues were highest in dust soil from homes located in closest proximity to agricultural operations.[14] In California, the children of migrant farmworkers living near sprayed fields experienced depressed cholinesterase activity and symptoms of acute pesticide exposure. Nearly one in five of these children had below-normal cholinesterase levels even though they did not work in the fields. Residential exposure to pesticide drift was considered responsible.[15]

Carpets act as long-term reservoirs for pesticides that are sprayed indoors. A study assessing pesticide exposure from carpet dust in homes revealed that the average number of pesticides found in the carpet dust samples was twelve, compared to 7.5 in air samples collected in the same residences. Moreover, in all residences sampled, thirteen pesticides that were not detected in the air were found in the carpet dust. The neurotoxic insecticide diazinon was detected in nine of eleven carpets tested. Exposure may be further exacerbated when carpets are cleaned, allowing pesticides to become airborne again and available for inhalation.[16]

Slower Breakdown Rate

Pesticides applied to soil, water, vegetation, or other surfaces indoors usually break down at a slower rate than pesticides applied outdoors. This is due primarily to the lack of sunlight indoors. This includes glass greenhouses, as the glass filters out ultraviolet light necessary for pesticide degradation. Pesticides applied indoors are not affected by wind or rain, and are less likely to move by mass transfer from the point of application. Vapor loss may also be less, as surfaces are not exposed to the heat of the sun.

Pet Exposure

Children who play with pets treated for fleas, ticks, and other pests can be exposed to pesticides. Flea collars, shampoos, soaps, sprays, dusts, powders, and dips usually contain an insecticide. Common insecticides for pets include pyrethrins such as permethrin and OPs such as chlorpyrifos, diazinon, and phosmet.[17] A study of 238 households in Missouri found that 50 percent used insecticides to control fleas and ticks on pets.[18]

Pesticides in Farmworkers' Homes

Children of farmworkers can be exposed to pesticides through multiple pathways, including agricultural take-home and drift as well as residential applications. Because farmworker families often live in poor-quality housing, the exposure from residential pesticide use may be substantial. Eight locally reported agricultural pesticides and thirteen pesticides commonly found in the homes of forty-one farmworker families with at least one child less than seven years of age in western North Carolina and Virginia were measured. Wipe samples were taken from floor surfaces, toys, and children's hands. Results indicated that six agricultural and eleven residential pesticides were found in the homes, with agricultural, residential, or both pesticides present in 95 percent of homes sampled. In general, residential pesticides were more commonly found. The presence of both types of pesticides on floors was positively associated with detection on toys or hands. Agricultural pesticide detection was associated with housing adjacent to agricultural fields. Residential pesticide detection was associated with houses judged difficult to clean.[19]

Dangers of Using Farm Pesticides in Homes

Some pesticides are formulated differently for farm use and homeowner use, yet they often have the same trade name. Examples include 2,4-D, Roundup, and diazinon. Homeowners may be tempted to use a small amount of a farm-formulated product in the home, thinking that it is the same material as the homeowner product, only cheaper. Although the active ingredient may be the same, the two products are different. Pesticides formulated for use on crop pests include inert ingredients to help carry the product into crop areas. These inert ingredients can cause problems on some vegetation growing in a home environment, and can present health risks for people who have excessive contact with the material applied around the home. Typically, farm pesticides are more concentrated than home products with the same active ingredients. They haven't been tested for persistence, potential for damage to plants likely to be grown around the home, or for home applicator safety. Using any pesticide in a situation that is not provided for on its label is illegal.[20]

Termite Control Hazards

The most common, unseen, and hidden poisons are pesticides used for termite control (before and after construction) or interior pest control, and pesticides used in gardens or on lawns where contaminated soil can be tracked into the home. Many termiticides could have been applied a decade or more earlier. Frequently, a new crack in the foundation, recent water damage, a recent termite infestation, or a termite inspection with repeated treatment can cause the release of a termite pesticide from the soil or holes into the indoor environment of the home. There may have been a misapplication of the pesticide, such as using too much or spilling it. A family may move into such a treated house without any information or records provided by the seller, the bank, the real estate agent, or the real estate attorney. It should be noted that revealing information about termite applications is not required by law, but is sometimes by banks and financial institutions to protect their claim on the property.[21]

Factors Involved in Health Risks

Since the main ingredients in pesticides can be organic, they can affect vision and memory. Health effects resulting from pesticide exposure are dependent on specific products and formulations. General-use pesticides available for homeowner use indoors are usually aerosols (spray cans and foggers), ready-to-use (pumps and liquids), pet products (flea and tick shampoos for dogs and cats), and baits (rat and mouse poisons). In order for a toxic effect to occur, direct contact by mouth, skin, or lungs must occur. Specific "dos" and "don'ts" for a pesticide product are on the label under the precautionary statement section. The best protection from exposure is to read and follow the label.

Irritation to the eyes, nose, and throat can occur with the use of aerosols and foggers if ventilation directions are not followed. In addition, disinfection of bathrooms, especially toilets, can also result in irritation. Overusing too much of one product, using one product too often, or using several products at the same time may also cause overexposure.[22]

Risks to Children

Pesticides sold for household use, notably impregnated strips and foggers, or "bombs," which are technically classed as semivolatile organic compounds, include a variety of chemicals in various forms. Exposure to pesticides may cause harm if they are used improperly. However, exposure to pesticides via inhalation of spray mists may occur during normal use. Exposure, particularly to children who may be in close contact with contaminated surfaces, can also occur via inhalation of vapors and contaminated dusts after use.

Increased odds of childhood leukemia, brain cancer, and soft tissue sarcoma have been associated with children living in households where pesticides are used. Other childhood malignancies associated with pesticide exposures include neuroblastoma, Wilms' tumor, Ewing's sarcoma, non-Hodgkin's lymphoma, and other cancers.[23]

Home and Garden Pesticide Survey

The Research Triangle Institute of North Carolina conducted a one-time national survey in 1990 of more than 2,000 households in fifty-eight counties across the country, examining what pesticides are used for specific pest problems, how often they are used, how they are applied, how unused pesticides are stored or disposed of, how empty pesticide containers are disposed of, the extent of child-resistant packaging, the effectiveness of pesticides, and which pests are major problems. The sample was representative of an estimated 84,573,000 households. The findings of the survey included:

1. About 85 percent of the households surveyed stored at least one pesticide product when the survey was conducted. Most, about 63 percent, had between one and five pesticide products in the home.

2. Fire ants were the most serious problem of the surveyed households, with cockroaches, other types of ants, fleas, and mice or rats being the most serious problems inside the home.

3. The most common means of disposing of empty pesticide containers and left-over pesticides was to include these items in household trash.

4. Many households still had pesticides whose registrations have been cancelled by the EPA. Products containing chlordane were calculated to be present in one million households; DDT-containing products in 150,000 households; heptachlor in 70,000 households; and silvex in 85,000 households.

5. Only about 20 percent of all pesticides in the home were found to be stored in child-resistant packaging.[24]

Homeowner Awareness of Pesticide Risks

There is little public awareness and understanding of pesticide risks, as evidenced by the ever-increasing use of pesticides on home lawns and gardens. The pursuit for the perfect yard has driven many homeowners to increasingly use readily available, heavily advertised pesticides, including "weed-and-feed" pesticides.

Risk reduction is an important factor for people who use pesticides in the home and outdoors. Only a small percentage of the public is required to demonstrate competence with regard to pesticide use for private and commercial applications. Many homeowners use pesticides every day but are not required to possess pesticide knowledge before buying and using many products. This lack of formal training can result in overuse of pesticides, inappropriate use, overexposure, and increased waste-pesticide disposal concerns.

Homeowners rarely read the complex instructions on product labels. While treatments against some pests are important, most people have little information to help them decide when to put others and the wider environment at risk. As much as 20 percent of pesticides are disposed of by householders pouring them down the drain. This can pollute rivers and drinking water. It only takes one tablespoon of some pesticide concentrates to breach drinking-water standards for 200,000 people.

Meanwhile, unused pesticides linger in garden sheds and kitchen cupboards for years, presenting hazards to children, pets, and wildlife. Disposing of pesticides accumulated over several years with ordinary household refuse can pose serious risks to the environment and public health. Ordinary landfills are not designed to accept hazardous waste.[25]

Understanding Pesticide Disposal

Pesticides are often relegated to storage shelves because they are difficult to mix and apply, because they are not suitable for the task at hand, or because too much product was purchased. When pesticide use is a necessity, consumers should buy ready-to-use products or concentrates that can be used up within a short period of time.

Unfortunately, some households dispose of leftover pesticide products by pouring them into the sink, toilet, street, gutter, sewer, or onto the ground. Such disposal "sites" are unacceptable. Household pesticide product labels generally will indicate that partially filled containers may be wrapped in several layers of newspaper and discarded in the outdoor trash. But for many people this disposal option is neither acceptable nor environmentally sound. Unused pesticides are best disposed of by using the products on the sites indicated on the label. Additionally, empty containers should be discarded in the household trash so that they are not reused.[26]

The Ten Most Frequently Used Household Pesticides and Their Effects

The ten most frequently used household pesticides are:

1. 2,4-D is the most popular, with nearly 40 million pounds a year being used in more than 1,500 different herbicide products in the United States. It is known to cause lymphomas and various cancers.

2. Diazinon is used for ant and roach sprays, and is highly toxic to birds and fish. None of the studies performed for its registration are considered adequate. Millions of birds are dying each year because of diazinon and other lawn chemicals.

3. Carbaryl (used, for example, in flea and tick powders) kills honeybees and causes birth defects in dogs. No studies of this compound are considered adequate.

4. Methoxychlor (for example, in insect sprays, cat flea powders, and fruit tree sprays) does not have toxicity data up to current standards.

5. Chlordane is a known carcinogen and is no longer being manufactured, but existing supplies are still in use.

6. Chlorpyrifos, used against termites and fleas, is extremely toxic to animals, and human exposure may exceed recommended limits due to use in both agriculture and homes. The EPA's acceptable toxicity levels only consider food sources, but exposure in many households is well above these levels when all sources are considered.

7. Malathion is widely used in home products for roaches, as well as in gardens for orchards and roses.

8. Maneb is a fungicide used in garden products, despite a complete lack of data on home exposure, and incomplete data on the chemical.

9. Simazine is an herbicide used in pools and ponds to kill algae. Not enough is known about the effects of swimming in water treated with this compound.

10. Captan is a fungicide used on tomatoes, vegetables, and other garden crops. It is used on roses, fruit trees, and even in shower curtains, paints, institutional bedding, and food packaging. It is a known carcinogen in animals.[27]

The EPA does not even know if two-thirds of the top fifty household pesticide ingredients may cause cancer or not.

More Disquieting Information

Dichlorvos (DDVP), which has been used in at least 12 million homes in this country, is now classified as a probable human carcinogen. It is emitted as a toxic vapor for three months or more from no-pest strips and flea collars. Pentachlorophenol causes blind spots in vision, corneal damage, and numbness, as well as problems with the autonomic nervous system. Chlordecone (Kepone) causes tremors and nervousness. Paraquat causes tremors and mental disturbances. Dieldrin and aldrin are organochlorines that cause convulsions, loss of coordination, and blurred and double vision. Lindane (in Kwell), another organochlorine like DDT, aldrin, endrin, and heptachlor, is used to treat head lice in more than 3 million children each year in the United States. It is known to cause cancer and stillbirths in animals, and it also causes symptoms ranging from headaches to convulsion. The other organochlorines have been banned because they accumulate in the body and cause tumors. Less-toxic alternatives to Kwell include Tripe-X, A-200 Pyrinate, and Rid, all of which contain pyrethrins, which are natural pesticides found in chrysanthemums. DEET is the most popular ingredient in insect repellents, and has been used by about 38 percent of Americans. When it is applied to clothing or skin, it is absorbed into the body. Documented symptoms of toxicity include slurred speech, staggering gait, agitation, tremors, convulsions, and death. Methyl bromide used in fumigation can cause drowsiness and double vision.

In all, well over a million Americans are estimated to be at risk of pesticide toxicity with damage to the brain, nerves, eyes, lungs, liver, kidney, and endocrine glands.[28]

The Chlordane Problem

Chlordane was a pesticide used to prevent or eliminate termites during the 1950s, '60s, '70s, and '80s. However, after many reports of serious illness among both adults and children following its application, and evidence linking it to cancer in animals, chlordane was finally banned by the EPA in March of 1988. Unfortunately, the ban did not occur until more than 30 million homes throughout the United States had been treated. Concerns in Florida were even greater because of the increased termite problem and the fact that research shows chlordane is higher in homes built on sandy soils.

Most homeowners are unaware that just before the concrete slab was poured for their home's foundation, a pesticide company saturated the soil with 100 gallons of chlordane per 1,000 square feet of area. People were literally building their homes on top of a toxic chemical dump. The public was reassured by the pesticide industry and entomologists that this was a safe procedure. It was thought that the chemical would not enter into the home because of the barrier provided by the cement foundation. However, this turned out not to be the case.[29]

Decades-Long Chlordane Contamination Results

Chlordane is such a highly toxic and persistent chemical that homes treated decades ago are still showing unsafe levels of it in indoor air. Problems develop because the hundreds of gallons of chlordane applied underneath the home are entering it, slowly evaporating and rising through cracks in the foundation or around plumbing pipes. This became evident in the 1970s when the U.S. Air Force randomly tested more than 500 apartments and housing units of its airmen. Results showed approximately 75 percent of the units tested contained chlordane in the air and more than 5 percent were above safety guidelines of five micrograms per cubic meter of air.

Unfortunately, this has turned out not to be an isolated case. Studies by the New Jersey Department of Environmental Regulation and other agencies have found similar results in hundreds of homes in New Jersey and New York. Of great concern, when testing sixty-four homes built before 1980, researchers found more than 30 percent of the homes contained chlordane levels above the five microgram safety limit set by the National Academy of Sciences.[30] There are now several university studies showing that even so-called acceptable levels of chlordane in indoor air can cause respiratory and neurological problems. These are discussed below.

Illnesses Linked to Chlordane Home Exposure

A study of 261 people who were either living or had lived in homes with higher air chlordane levels were found to have nearly three times more respiratory illnesses, including sinusitis and bronchitis. The study, conducted by the School of Public Health at the University of Illinois, also found other illnesses significantly more often in the chlordane homes. These included chronic cough, anemia, neuritis, ovarian/ uterine disease, and skin disorders. Migraine headaches, the worst acute symptom found, were occurring in 22 percent of people living in the higher-level chlordane homes.[31]

According to recent statistics, one in eight women in the United States will develop breast cancer, with rates nationally three to seven times higher than those in Asia. A 2005 study conducted by the U.S. Army Institute of Surgical Research and Texas Tech University Health Science Center in Lubbock, Texas, revealed that cancerous human breast tissue contained the chemical heptachlor epoxide at levels four times higher than in non-cancerous breast tissue (heptachlor epoxide is found in chlordane). An estimated 50 million Americans are currently exposed to the volatilization of this chemical from previously treated pre-1989 homes.[32]

An excellent test to determine how well a person's immune system is functioning is called proliterative response. This test measures how fast a person's immune system cells multiply in order to eliminate invading bacteria or viruses. In several different tests of proliterative response, conducted at the Southern Illinois School of Medicine, it was found that people living in chlordane-treated homes had immune system cells that multiplied only about half as fast as the immune system cells of people not exposed to chlordane. In another immune system test conducted by the same scientists, eleven of twelve chlordane-exposed persons were found to have a condition known as autoimmunity. This is where the individual's own immune system mistakenly attacks his or her own body, which can result in a variety of other illnesses.[33]

Another unexpected symptom of chlordane exposure is an increase in body weight. In fact, in an experiment of more than twenty test animals receiving chlordane exposure equal to that sometimes found among the U.S. population, there was an average 8 percent increase in body weight. The body weights of animals who received 500 nanograms of chlordane increased by an average of 11 percent. Chlordane exposure has been shown to reduce by half the levels of some hormones in female test animals; however, scientists are unsure if this is the actual reason for the weight gain being observed or if it is due to another reason, such as changes in the areas of the brain which control body weight. This raises the question of whether the same symptom may be occurring among residents living in chlordane homes built before March 1988.[34]

Illegal Application of Pesticides

Individuals should always hire only pest control operators licensed or certified for commercial and/or residential structure application to treat homes for pest control. People should always ask to see licenses, certifications, and picture identification cards, and make sure they are current. These individuals should be insured and should provide written proof of insurance. The company's employees must always be bonded, which means that the company is responsible for reimbursing homeowners for any loss or damage caused by its workers. A licensed pest control operator also must always provide the label and labeling information of any pesticide products that will be applied in or around a home. Homeowners should always read the entire label to ensure that it contains EPA establishment and registration numbers, and make sure it is used and applied strictly according to the label or labeling. Additionally, homeowners should never purchase pesticide products from other than established retail businesses. Pesticide products should only be sold in unopened, original pesticide containers that are fully labeled and contain an EPA registration number. Another very important reminder to homeowners is to always store pesticide products in locked cabinets out of the reach of children and pets. Pesticide products should also be disposed of according to label instructions.[35]

Incomplete Label Data

Label information addresses acute or immediate effects only. Information about chronic or long-term hazards of chemical products, such as cancer or birth defects, is not provided to purchasers. For example, some inert ingredients are toxic, but only the percentages of inert ingredients are required on the label, not their identities. Another frustration for homeowners is the fact that many chemicals have numerous and/or scientific names that make it difficult to compare products. Antidotes listed on the label in case of poisoning may be incomplete, out-of-date, or even dangerously wrong. Also, many labels do not indicate how to dispose of a product safely. The use of the term "non-toxic" is for advertising only. It has no regulatory definition by the federal government.

Store Employees Untrained Regarding Pesticide Use

The majority of homeowners purchase pesticides from home and garden centers and use these outlets as information for pest management recommendations.[36] A troublesome finding from a statewide survey in Illinois indicated that only 34 percent of retail stores surveyed provided any employee training related to pesticide use. Furthermore, of those individuals who received any training, only about one-half stated that the level of training was adequate. If any training was provided, the focus tended to be on pesticide selection and use, while the concepts of IPM were largely ignored. Because many stores hire seasonal employees during the spring and summer months, there tends to be a high turnover of employees. Regular training programs are important to educate new employees in pesticide use.[37]

Pets and Pesticides

Many household pesticides carry warnings on their labels cautioning people to keep pets away from treated areas. In the case of flea powders and other pet pesticides, to avoid pet illness, warning labels usually state application rates and the minimum age of any pet to be treated. As is the case for warning about human health hazards, these guidelines cannot assure that your pet will suffer no adverse effects. Pesticides are known to poison fish and other forms of wildlife when used outdoors; a similar health hazard exists for aquarium fish and pet birds during and after pesticide application.

In recent years, hundreds if not thousands of pets have been poisoned. Products containing OPs are among the worst culprits. The EPA finds that these pet products are frequently misused and that manufacturers should anticipate this. Cats are particularly vulnerable, since they often lack key enzymes for metabolizing or detoxifying OPs. As with children, a cat's small size and unique behavior—in this case, grooming—work against them as well, making them particularly vulnerable to OP poisoning.[38]

In a Lighter Vein: Reckless Endangerment

In California, the Department of Pesticide Regulation (DPR) has compiled a list of what might be called "The Top Ten Pesticide Blunders at Home." It is reminiscent of the immensely popular television show of yesteryear, *Candid Camera*, but the important difference is that no one's life was jeopardized by Allen Funt's histrionic efforts. These illustrations of present-day human foibles could have had disastrous consequences. Fortunately, not one fatality was involved, although most victims required medical treatment.

Most incidents occurred in 2004 and 2005 and were compiled by the DPR's Pesticide Illness Surveillance Program. The Top Ten in no particular order follow:

1. A San Diego County man preparing to spray ants with insecticide failed to notice the aerosol can faced the wrong way. He sprayed himself in the face, developed respiratory symptoms, and sought medical attention the next morning.

2. In Los Angeles County, a woman sprayed an aerosol insecticide under her kitchen sink to kill roaches. To get a better shot, she stuck her head inside the cabinet and then inhaled fumes. Her lungs began to burn and she sought medical attention.

3. An Orange County resident set off two bug bombs and left his house. He returned 90 minutes later, opened the windows, and remained inside. He developed heart symptoms and went to a hospital, where he suffered a stroke.

4. Another Los Angeles resident who sprayed her kitchen to kill flies drank from a glass of water that sat uncovered in the same room while she sprayed. A runny nose, headache, and chest tightness prompted her to seek medical aid.

5. In Orange County, a dog owner with asthma hugged her one-pound puppy shortly after it received a liquid flea control treatment from the woman's veterinarian. It was later determined that the puppy was treated with a dosage meant for larger dogs. The owner experienced shortness of breath, blurry vision, and other symptoms. The puppy also apparently suffered ill effects.

6. A San Diego receptionist sprayed an insecticide around doors in her office for spiders. She got the pesticide on her hands so she rubbed them together. She later rubbed her eyes. Her hands and eyes began to itch, so she sought medical attention.

7. A San Bernardino truck driver prepared to disinfect his tires with a hose-mounted sprayer. When he pulled on the hose, it knocked the attached disinfectant bottle off. The bottle hit the ground and disinfectant splashed into his face and eyes.

8. A Los Angeles County worker prepared to mop a kitchen floor when she noticed she was almost out of the usual cleaning product. She mixed bleach with the cleaning product, which created fumes. She developed respiratory symptoms and sought medical attention.

9. At a San Bernardino County fast-food outlet, a customer at the drive-through window bought iced tea and noticed a foul taste, followed by a burning throat and nasal passages. The cup apparently contained some sanitizer from

an improperly rinsed tea machine (a similar case was also reported in Los Angeles County).

10. A Marin County lifeguard mistakenly added muriatic acid to a chlorine tank. He inhaled the resulting fumes and developed symptoms. His mother saw him coughing and took him for medical aid.

These blunders graphically demonstrate what NOT to do as you undertake household and gardening chores or other work with pesticides. Health and safety scientists say a few simple precautions can prevent most pesticide accidents:

- Look for the least-toxic solution to pest problems, both indoors and out.

- Read all pesticide label directions closely and follow the directions to the letter.

- Keep pesticides in their original containers and out of children's reach.

Many home pesticide accidents occur in kitchens and bathrooms, although they often go unreported. Children are especially vulnerable when adults put pesticides into drinking containers, such as soda or juice bottles. Consumer pesticide products with colorful packaging and attractive scents may also attract children.[39]

The Good Guys Win—Sometimes

When George and Carolyn Fox called the Orkin extermination company to spray their house for termites in 1993, little did they know the trouble that would ensue. The exterminator used existing stocks of the chemical chlordane, which had been banned in 1988, and contaminated the couple's home to the point that it was uninhabitable. The couple sued the company, and on November 20, 1998, won the suit, receiving almost $2 million in damages. According to the *Tampa Tribune*, they were awarded $200,000 for their historic home, $200,000 for mental anguish, $1.2 million in punitive damages, and $168,000 to replace the values of antiques that Mrs. Fox collected. The house was also condemned by state health officials.[40]

Final Thoughts

Almost every household uses pesticides. But most people do not understand that pesticides can be dangerous. Bug spray, flea powder, rat poison, and garden weed killer are all types of pesticides. These products contain chemicals that kill pests. That also means they can harm humans if they are not used safely.

Parents can eliminate the use of pesticides in and around their homes by using least-toxic pest control methods, excluding pests by caulking cracks, and keeping kitchens and other parts of the home free from food sources that attract pests. Low-toxicity self-contained baits should be utilized instead of spraying potent toxicants directly into the home and environment.

If blame is to be assessed, it must be placed squarely on the shoulders of the U.S. regulatory system, which allows dangerous chemicals to be put into consumer products, does not require even minimal safety testing for the majority of chemicals currently in use, and has virtually no prohibitions in place to reduce exposure to chemicals known to cause harm. The U.S. chemical industry must also take responsibility for failing to replace chemicals of known toxicity with safer substitutes.

The American people deserve to be safe in their own homes, and should be able to purchase and use products without unwittingly exposing themselves and their children to substances that can cause cancer and disrupt development. There is solid evidence that the federal government, the states, and the pesticide industry must take immediate action to replace harmful chemicals with safe substitutes.

Lawn and Garden Pesticides

Patricia J. Wood, Executive Director of Grassroots Environmental Education, has stated: "We are in the midst of a revolution in scientific understanding of the links between environmental exposures and health, and pesticides appear to present significant risks. With an abundance of safe pest controls available today, the routine use of lawn care pesticides should not occur."[41]

Background

Lawns are a standard feature of ornamental private and public gardens and landscapes in much of the world today. Lawns are created for aesthetic use in gardens and for recreational use, including sports. They are typically planted near homes, often as part of gardens, and are also used in other ornamental landscapes and gardens.

Americans keep lawns to provide themselves and their families with a cool green oasis where they can play and relax. But dousing lawns with toxic pesticides means that a space meant for families and pets can become a potent danger to health and the environment.

Americans maintain over 32 million acres of lawns, with an average amount of five to ten pounds of pesticides (counting active ingredients) used per acre per year. That rate of pesticide use is more than three times higher than pesticide use on farms. It means that we are exposing our children, our environment, and ourselves to the unintended effects of as much as 200 million pounds of pesticides nationwide from lawn care alone.[42]

Health Dangers

Why should there be concern about this astronomical level of pesticide use? Obviously, reduced use of lawn and garden care pesticides could prevent many of the thousands of pesticide poisoning that occur annually. But perhaps more importantly, many of the pesticides in use today are associated with long-term human health problems.

According to the EPA, the majority of lawn care chemicals in use today are possible or probable carcinogens. A National Cancer Institute Study indicated that children in homes where lawn and garden pesticides were used were 6.5 times more likely to develop acute lymphoblastic leukemia than those living where pesticides were not used.

Despite labels that indicate that treated lawns are safe for human and animal contact after twenty-four to forty-eight hours, many lawn and garden pesticides have been found to persist far longer than that period of time. As previously mentioned, pesticides can also contaminate indoor environments when they are tracked or blown inside. Thus family members and pets can be exposed to pesticides when playing on a treated lawn even after the reentry period listed on the label.

Even when used as directed, pesticides can kill non-target organisms, such as beneficial insects, desirable plants, birds, and other wildlife, along with their target pests. Because most pesticides work by interfering with physiological processes shared by many organisms, they can kill indiscriminately. Crashes in honeybee populations, damage to wild plant life, fish and bird die-offs have all been linked to herbicide and insecticide use.[43]

Of the commonly used lawn pesticides, nineteen are carcinogens, thirteen are linked to birth defects, twenty-one are linked to reproductive effects, fifteen are neurotoxic, twenty-six may cause liver or kidney damage, twenty-seven are irritants, and eleven can disrupt the hormone system. Pregnant women, infants, children, the aged, and the chronically ill are at greatest risk from pesticide exposure. Pets too are regularly poisoned.

A report by the Toxics Action Center on the company ChemLawn, the largest provider of lawn care services in the United States, uncovered that more than 40 percent of the chemicals in ChemLawn's consumer product range contain ingredients banned in other countries. All of the products in their lineup not only pose a threat to human health, but to water supplies, aquatic organisms, and non-target insects.[44]

Impact on Water Supplies

Lawns and gardens treated with pesticides and fertilizers can be a significant source of surface water contamination when the chemicals used run off into neighboring water bodies. When pesticide residues contaminate waters, they can kill small plants and animals at the bottom of the food chain as well as damaging some fish species.

Homeowners may unknowingly contaminate their own well water by using pesticides on their lawns. Factors that influence a pesticide's potential to contaminate water include physical and chemical factors, environmental influences, application methods, and other practices associated with the pesticide use. Only two of the top five lawn care pesticides, 2,4-D and glyphosate, are regulated under the Safe Drinking Water Act, despite governmental acknowledgment of the intensity of the effects of their release on the environment, and their potential to leach into groundwater supplies.[45]

False Assumptions

Consumers presume that lawn care pesticides are safe because they are sold in stores that also market foods and other consumer products. Products with names such as Weed-and-Feed, Weed-B-Gon, and Turf Builder with Plus 2 Weed Control might seem innocuous to the consumer, but they contain pesticides such as 2,4-D and MCPP, which have been associated with soft tissue cancers. Products such as Bug-B-Gone and Turf Builder with Insect Control also might sound quite benign to the consumer, but they contain carbaryl and diazinon, both of which are capable of harming the nervous system. Carbaryl is suspected of altering human hormone function, while the residential uses of diazinon pose a special threat to children. The public remains uninformed of the potential health threats posed by these chemicals, while at the same time being subjected to intensive television, radio, and Internet advertising.[46]

Several Particularly Dangerous Lawn and Garden Pesticides

A wide variety of insecticides are available for use in gardens, but many contaminate water and soil, kill beneficial insects, and can harm health. Many lawn and garden pesticides are toxic chemicals that should not be used around children, sensitive individuals, or pets. These include the following:

Carbaryl is the main ingredient in Sevin dust and Bonide spray, and is associated with a stunning array of human health problems. It contaminates groundwater and is toxic to many kinds of wildlife, killing as many as 1 to 2 million birds in the United States every year.[47]

Malathion, often used for control of adult mosquitoes, is toxic to the human nervous system, like all OP insecticides.[48]

MCPA is a common ingredient in products such as Trimec (along with MCPP and dicamba) and weed-and-feed products like Scott's Pro Lawn; it is frequently used by lawn care companies. It is toxic to wildlife and humans and a possible carcinogen.[49]

The herbicide 2,4-D is sold under a variety of brand names and is one of the most widely used herbicides in the world. Americans use 9 million pounds of it every year to control lawn and garden weeds. Long-term exposure has been linked to damage to the liver, kidneys, and the digestive, muscular, and nervous systems, and may also be linked to non-Hodgkin's lymphoma.

Glyphosate is the second most commonly used home and garden herbicide, with 25 million applications annually. If inhaled, glyphosate can cause respiratory problems, nose and throat irritation, lung congestion, and an increased breathing rate.[50]

Lawn and Garden Pesticides Increase Indoor Risks

Before going indoors, people should wipe their shoes carefully after walking on a lawn or garden treated with herbicides; otherwise they could be tracking in dangerous chemicals that will linger in carpets and other parts of the household for a year or more. Researchers found that 3 percent of "dislodgeable turf residues"—the portion of a pesticide that does not adhere to the turf—were tracked indoors. In homes with carpeting, almost all of the pesticide became deeply imbedded carpet dust, where it couldn't degrade through exposure to sunlight, wind, rain, or soil microbes. Although the investigation revealed that only 10 percent of the residue remained on carpet surfaces where it could easily contact human skin, previous research has suggested that transport of pesticides from lawns and gardens presents significant risks of human exposure, particularly for children.

In the study, researchers applied a pesticide formulation containing 2,4-D, dicamba, and mecoprop (X-Gro Broadleaf Weed Killer) to sections of a lawn that had not been treated with pesticides for at least ten years. Participants then walked on the treated areas, staggering their times and walking in different areas so that most of the treated ground was covered. They then either wiped their feet on a mat or walked directly onto indoor carpeting, both of which had never been used before. Researchers analyzed residues tracked onto the carpets as well as levels of dislodgeable turf residues on the lawn.

Use of entryway mats reduced the level of pesticide residues on carpet surfaces by 25 percent and reduced carpet dust residues by 33 percent. Estimates were that 2,4-D residues could remain in household carpet dust for up to one year after small turf applications.[51]

Another study found that up to 0.2 percent of the residues of two different herbicides applied to a lawn were dislodgeable. Notably, the amount of residue that was dislodgeable actually increased between four and eight hours after the application, as the pesticide spray dried.[52]

Another investigation revealed that 1.5 to 4 percent of residues of the insecticide chlorpyrifos deposited on a lawn could come off the treated lawn onto shoes, skin, or clothing.[53]

Home Garden vs. Commercial Use Pesticides

Pesticides for home garden use are not necessarily of low toxicity. Active ingredients available to the gardener can be extremely toxic. The technical products of strychnine (LD_{50} = 30–60 mg/kg) and Di-Syston (LD_{50} = 4 mg/kg) are readily available at nurseries and garden centers. Anyone can buy them. The assumption that the commercial applicator may use pesticides with more toxic active ingredients than the gardener is wrong. Remember, it's the dose that makes the poison.

Because of the small label size, home garden products may not list all of the plants and/or pests for which the product may be registered for use. For example, one bottle of ACME 25 percent EC, may be called ACME Fruit and Vegetable Insect Control, while another bottle of ACME 25 percent EC may be called ACME Insect Spray. Both may be basically the same product, but the plants and pests listed on the label can vary greatly. This situation causes some confusion in pesticide applications and encourages the purchase of excessive amounts of pesticides. It is not legal to tell anyone that they can use a pesticide on a crop or site not listed on the label even though another similar product may have the crop listed on the label.

Products packaged for the commercial grower may appear to be less expensive, but homeowners should not be tempted to use them. They are generally more concentrated than those for home use and require special protective clothing and equipment for application. These products are in larger containers than the homeowner could expect to use or store safely, and are much more difficult to calibrate and mix correctly since rates are usually based on a per-acre system.

In the majority of cases, the site listed on the label is a field crop and not a lawn or garden, so any use of it on a lawn or garden would then be illegal. A few products that are extremely toxic to humans or the environment are classified by the EPA as restricted-use pesticides. The label will state "restricted-use pesticides for retail sale to and application only by certified applicators, or persons under their direct supervision." Certification from a state department of agriculture is required by law for purchase and use of restricted-use pesticides. This certification is intended for commercial and private applicators (farmers) and does not automatically allow the use of these products by home gardeners.[54]

Sales, Labeling, and Packaging

A survey of eighteen stores in Connecticut found that most stores displayed pesticide packages with visible tears or rips. Their contents had visibly contaminated store shelves, floors, and storage areas.

The packaging of many lawn care chemicals is porous, releasing vapors from the chemicals into the nearby air. These vapors are easily detected by sense of smell, and often contaminate indoor air where sold.

Pesticide labels do not provide the consumer with sufficient warning and instruction regarding the toxicity of contents, pesticide potential to contaminate water supplies, effects on fish and wildlife, and proper handling and disposal. Labels on the front of packages claim product benefits in multicolored letters often several inches high, while warning information and directions for safe use and disposal are commonly displayed in small type on the back.

Some lawn and garden packages require the removal of a plastic wrapping to access multi-paged warnings about product ingredients, often printed in minute type. These lawn and garden pesticides are commonly sold in stores that also sell food and other consumer products that are vulnerable to chemical spills.[55]

Beware Fertilizer-Pesticide Combinations

Many homeowners and lawn care companies routinely combine fertilizer and pesticides in a series of applications throughout the spring, summer, and fall. These multi-step programs are promoted as the sure and easy path to the perfect lawn. The pressure to have a perfect lawn, however, has clouded a number of issues and literally mixed ingredients that should be kept separate. Areas of caution include:

1. **Routine insecticide applications.** Most insects found on a lawn are beneficial, and insecticides can harm them. Research in Wisconsin indicates that only about one lawn in 200 will need an insecticide application in a given year. Even on lawns where harmful insects exist, natural controls or better lawn care practices will reduce the threat. For example, chinch bugs can be pests during a dry year, but proper watering (or even a good rain) can minimize their effects.

2. **Routine herbicide applications.** Weeds are not the cause of an unhealthy lawn; they are the result. The best defense against weeds is a thick, healthy lawn that can be attained via proper watering, fertilizing, and mowing. Routine herbicide applications are unnecessary and their effects can be misleading. For example, weed-and-feed products are widely used to kill dandelions in spring, when the flowers are so noticeable. The curling weeds seem to indicate that the herbicide has been effective, but in fact the herbicide may kill only the top of the weed, not the root.

3. **Unnecessary nutrient applications.** Most commercial fertilizers contain phosphorus, a major water pollutant. Yet many soils already contain enough phosphorus for a healthy lawn. This underscores the need for a soil test before applying fertilizers. Low-phosphorus or phosphorus-free fertilizers can provide nutrients while avoiding the threat to water quality. In short, applying unneeded pesticides and nutrients in a generic, multi-step fertilizer program can be expensive for the homeowner and harmful to the environment.[56]

Follow the Label

Consumer awareness of pesticide use should be given a high priority. Homeowners should be encouraged to obtain information on the pesticides that are used on their lawns. Homeowners should be sure to read and follow pesticide labels carefully for any products they apply and should always ask to see the label of any products commercial lawn care services use before these products are applied.

However, roughly half of homeowners admit they don't read or follow label directions when applying pesticides to lawns, often using significantly more than the recommended amount, lawn care experts maintain.[57]

Lawn Pesticide Notification Laws

Notification

Notification of pesticide applications provides the public with the opportunity to take precautions to avoid direct exposure to hazardous pesticides. Twenty-one states have adopted laws requiring notification of lawn, turf, and ornamental pesticide applications by hired applicators. Concerns over the potential public exposure to these pesticides have led states to pass laws that warn neighbors of a lawn application by posting notification signs, establishing registries, or providing prior notification to abutting property owners. Because only 19 percent of U.S. households hire lawn professionals, some states also require that homeowners provide notification to those on neighboring properties. State notification laws usually indicate where, when, and what pesticide has been or will be applied and by whom. State notification requirements vary in specifics, but where prior notification is required, it generally provides notice twenty-four to forty-eight hours in advance.

Posting

Twenty states require that commercial applicators post notification signs when a pesticide is applied to a lawn. Most states require that notification signs be posted in a conspicuous point of access to the treated property and left in place for twenty-four hours. Warning signs vary in language, but usually state: "Lawn Care Application: Keep Off the Grass." In Connecticut, homeowners and commercial applicators are required to post notification signs if applications are made within 100 square feet of unfenced turf. Wisconsin retail stores are required to provide warning signs to homeowners when they purchase pesticides. The U.S. District Court in Seattle requires in-store notices to consumers on lawn chemicals and endangered salmon in West Coast states.

Registries

Thirteen states require that a state agency or, in some cases, individual companies, establish a registry for people to sign up for prior notification when an adjacent property is treated with a pesticide by a commercial applicator. Generally, the states with such laws include provisions that require an applicator to inform any person on the registry of an upcoming pesticide application to property adjacent to theirs. Some states, including Florida, Maryland, and Pennsylvania, have the additional requirement that individuals requesting notification provide documentation and certification from a physician. Registries only provide advance notice to those who make a prior request to be notified, and therefore are limited in providing adequate warning to the public.

State Preemption of Local Laws

Forty states preempt local ordinances on pesticides. However, two of those states, Minnesota and Montana, allow municipalities to adopt specific language regarding posting for commercial turf pesticide applications. Under New York state's lawn notification law, counties can adopt specific provisions that require commercial applicators to provide forty-eight hours' prior notice to all neighbors if treatment occurs within 150 feet of abutting property. It also requires homeowners to post notification signs of lawn application. Only six counties have set these requirements.[58]

Lawn Posting Problems

Posting of lawns has been a controversial practice. Any successful posting program needs to address the following problems:

1. Those most at risk, small children and pets, can't read.

2. The signs are often so small and low to the ground that it's difficult to read the message and still avoid contact with treated grass.

3. If signs are left to weather and rot, they would lose any effect they may have once had.

4. Some lawn pesticides have a very low degree of toxicity. To require posting of lawns treated with those pesticides may raise unwarranted alarm.

5. No one sign is suitable for all pesticides, yet multiple signs pose a logistical problem.

6. On a percentage basis, homeowners apply far more pesticides than commercial applicators, but requiring them to post on treated lawns has a number of practical problems.[59]

Practical Problems

1. Ensuring that statutory deadlines for release of all data are met by the responsible state agency.

2. Requiring state agencies to submit data electronically.

3. Making sure that all data is made available to the public.

4. Requiring farmers to report their pesticide use directly rather than through sales reports.

Lawn Care Pesticide Sickens Family

Brenda Jones believed the lawn care applicator from TruGreen ChemLawn when he told her that the chemicals he was going to use on her lawn were so safe that he didn't even need to wear a mask. According to Brenda, she was still hesitant and asked him to wait until she was safely inside the house before he started to spray. While Brenda waited for her dog, the applicator began spraying some fifteen feet behind her. Suddenly her eyes began to burn and a cough welled up in her throat. She turned to see a cloud of silver mist coming from the nozzle held by the applicator. Instantly, she grabbed the dog and dashed into the house to escape. She washed herself and the dog off, but it was too late; the damage had been done.

Being a registered nurse, Brenda was instantly aware of being sick. Her eyes, throat, and chest burned. Her head pounded like a drum. She became nauseous and coughed continuously. That evening her husband, Wayne, and two children, Jeffrey, age seven, and Kara, age three, were afflicted with headaches, dizziness, diarrhea, and other symptoms, causing Brenda great concern. They contacted the lawn care company the next day and discovered they had been exposed to the common weed killer atrazine and to the synthetic pyrethroid bug killer bifenthrin. Closing the windows did not prevent the pesticides from entering the house.

The following day the entire family was ill and even the dog, which had vomited the night before, did not move and would not eat—classic behavior for dogs who have been subjected to pesticide poisoning. Brenda called the family physician, and hoped the illness would pass on its own, a response similar to that of others who have been acutely poisoned.

Two days later, the doctor diagnosed her with chemical poisoning and prescribed antibiotics and steroids. Her children received the same treatment. When she contacted Florida's Poison Control Center to report the incident, they incorrectly responded that they did not handle pesticides. Later, a lung specialist advised Brenda that her airway had become reactive. With no cure in sight, her only option was to entirely avoid environmental chemicals.

Brenda had an impressive resume that described fifteen years of experience as a registered nurse at Johns Hopkins Medical Center, the University of California at Los Angeles, Stanford University, and the John Wayne Cancer Institute in California. Pesticides took away her livelihood. Future employment attempts failed as she became symptomatic with dizziness, weakness, and tremors, and she was unable to complete her shifts.

However, she states that her predicament is the least of her worries, as her son can no longer attend school due to reactions he suffers each time pesticides are applied to a field adjacent to the school. When she asked the applicator not to spray during school hours, he responded in the usual manner: "Weed killers and pesticides are registered with the EPA and are safe to use." "They won't hurt the children."

Brenda persisted, and approximately two weeks after the January 2003 incident, she contacted the Florida Department of Environmental Protection. In April, she received a letter from a division of the state's Agriculture Department. It indicated that the agency had questioned the lawn care firm, but that too much time had elapsed to attempt an on-site inspection to determine if a violation of the pesticide's label had taken place.[60]

Pesticide Education—Does It Work?

Education on the proper use of pesticides is often included in many lawn care and landscape management programs. Most often this is in the form of informational brochures or fact sheets on pesticide use around the home or garden. These information packets include tips on identifying pest problems and selecting treatment approaches that reduce environmental impacts, less-toxic pest control products if chemical control is necessary, and the proper mixing, application rates, and cleanup procedures for pesticide use.

The public perception that no alternative to pesticide use exists is probably the greatest limitation that these efforts encounter. Surveys tell us that the public has a reasonably good understanding of the potential environmental dangers of pesticides. Several surveys indicate that residents do understand environmental concerns about pesticides, and consistently rank them as the leading cause of pollution in their neighborhoods. Even so, pesticide use still remains high in many urban areas. The time required for homeowners to learn more about alternative pest control techniques may also limit educational efforts. Many residents prefer the ease of just spraying a chemical on their lawns to other pest control measures they perceive as more time intensive and less reliable.[61]

Organic Lawn Care—A Viable Alternative

Historically, organic lawn care has been practiced far longer than chemical lawn care. From the 1700s to World War II, lawns were maintained without chemicals. The lawns of eighteenth-century Europe differed dramatically from those that now cover suburban North America. Mass production of chemical pesticides and fertilizers during the 1940s ushered in the lawn's transformation from a collection of various grasses, legumes, and wildflowers to a sterile monoculture. Since then, the chemical lawn care industry has grown enormously. Many homeowners now believe that lawns require chemicals for survival. This simply is not true. A traditional lawn can be maintained without pesticides. In fact, widespread misuse of pesticides and fertilizers results in an unhealthy lawn that becomes chemically dependent and highly susceptible to pests and diseases.

The basic principle of organic lawn care is to nourish the soil. In this way, it differs fundamentally from chemical lawn care, which focuses on feeding the grass. Restoring and protecting soil health is crucial for the maintenance of a lush, naturally pest- and disease-resistant lawn. Healthy soil is alive. It contains a diverse collection of organisms, many too small to see with the naked eye, that interact in complex and intricate ways. Together, these organisms break down organic matter, make nutrients available for plant uptake, and aerate the soil. Soil aeration is important for water storage and air exchanges. Research shows that chemical fertilizers and pesticides degrade soil life and decrease the level of biological activity.[62]

Pesticide Manufacturers and Public Relations Efforts

At least seventy municipalities across Canada have restricted pesticide usage, including Toronto, Montreal, Vancouver, and Halifax, and the province of Quebec. Toronto's bylaw, for instance, applies to public lands as well as to private homeowners. In the United States, a coalition of twenty consumer and environmental groups launched a campaign that urges two of the largest home and garden retailers, Home Depot and Lowe's Home Improvement, to carry more organic lawn care products and to reconsider the sale of weed-and-feed products to protect the health of children, families, pets, and the environment.

This has fueled a high-stakes counterattack from the multibillion-dollar pesticide industry. Citing increased "activist threats" and recent pesticide bans in Canada and the United States, a nationwide coalition of pesticide manufacturers, suppliers, and lawn care companies recently launched a $1 million ad campaign to "educate consumers" about pesticides, according to a press release from the group, called Project Evergreen.

Environmentalists have responded, saying adequate testing has not been done on the vast majority of pesticides, and that health problems for humans, birds, and fish have been linked to lawn products.[63]

Realistic Considerations

Using pesticides for the goal of a perfectly manicured lawn, that is to say, for purely cosmetic reasons, involves unacceptable risks, exposing human life, pets, and wildlife to unjustifiable hazards. Living in the toxic stew of today's environment, it may not be possible to prove beyond a shadow of a doubt that lawn chemicals are at fault, but absence of proof is not proof of absence. We must err on the side of caution. No one's right to use lawn chemicals should take precedence over another's right to the highest level of health possible. If there is even a small chance that the use of lawn pesticides will contribute to a child developing leukemia or the exacerbation of an asthma attack, it's a simple no-brainer: don't use it.

Safe and effective alternatives for lawn care are available. The chemical and pesticide industries may claim that we need their products, but remember that these companies are in business, and their ultimate goal is profit, not health and well-being. Simple, safe, and inexpensive lawn and garden pest remedies could seriously harm the industry's bottom line. The time has come for industry to read the writing on the wall, and adjust their product lines and lawn care practices to those of a more environmentally friendly nature. One final idea to consider is that the heavy use of pesticides to create perfect lawns started only after World War II, and is an example of a socially accepted practice that needs to be reconsidered.

Notes

1. Representative Harold L. Volkmer, Democrat from Missouri, testimony before the Committee on Science and Technology, October 8, 1989.

2. J. D. Spengler and K. Sexton, "Indoor Air Pollution: A Public Health Perspective," *Science* 221 (4,605) (1983): 9–17.

3. Mary Beth Breckenridge, "Reduce Pesticide Use with Natural Alternatives," *Akron Beacon Journal*, May 13, 2006.

4. *Poison Prevention Tips* (Itasca, IL: National Safety Council, n.d.).

5. F. Immerman and J. L. Schaum, *Nonoccupational Pesticide Exposure Study (NOPE)* (Raleigh, NC: U.S. Environmental Protection Agency, Research Triangle Park, January 1990): 7–12.

6. Ibid.

7. R. A. Fenske et al., "Potential Exposure and Health Risks of Infants Following Indoor Residential Pesticide Applications," *American Journal of Public Health* 80 (6) (1990): 689–693.

8. R. J. Zwiener and C. M. Ginsberg, "Organophosphate and Carbamate Poisoning in Infants and Children," *Pediatrics* 81 (1) (January 1988): 121–126.

9. R. Lewis et al., "Evaluation of Methods for Monitoring the Potential Exposure of Small Children to Pesticides in the Residential Environment," *Archives of Environmental Contamination & Toxicology* 26 (1994): 37–46.

10. R. G. Lewis et al., "Determination of Routes of Exposure of Infants and Toddlers to Household Pesticides: A Pilot Study to Test Methods," presentation at the Air and Waste Management Association's 84th Annual Meeting, Vancouver, British Columbia, June 16–21, 1991: 8.

11. N. J. Simcox et al., "Pesticides in Household Dust and Soil: Exposure Pathways for Children of Agricultural Families," *Environmental Health Perspectives* 103 (12) (December 1995): 1,126.

12. Ibid.

13. M. Nishioka et al., "Measuring Transport of Lawn-Applied Herbicide Acids from Turf to Home: Correlation of Dislodgeable 2,4-D Turf Residues with Carpet Dust and Carpet Surface Residues," *Environmental Science and Technology* 30 (11) (1996): 3,313–3,320.

14. Simcox et al. Op. cit.

15. E. D. Richter, "Aerial Application and Spray Drift of Anticholinesterases: Protective Measures," in B. Ballentine and T. Aldrige, eds., *Clinical and Experimental Toxicology of Organophosphates and Carbamates* (Oxford: Butterworth-Heinemann Ltd, 1992).

16. R. Whitmore et al., "Nonoccupational Exposures to Pesticides for Residents of Two U.S. Cities," *Archives of Environmental Contamination and Toxicology* 26 (1993): 1–13.

17. M. Moses, *Designer Poisons* (San Francisco: Pesticide Education Center, 1995): 226–243.

18. J. Davis et al., *Family Pesticide Use in the Home, Garden, Orchard, and Yard*, study in partial fulfillment of a PhD degree from the University of California, Berkeley, 1991.

19. Sara A. Quandt et al., "Agricultural and Residential Pesticides in Wipe Samples from Farmworker Family Residences in North Carolina and Virginia," *Environmental Health Perspectives* 112 (3) (March 2004): 382–387.

20. Richard Pope and Sorrell Brown, *Why Do Farmers Use Pesticides?* (Ames, IA: Iowa Department of Agriculture and Land Stewardship, January 11, 2004).

21. Robert Simon, *Home Buyers Alert: What May Be in the Home You Are Buying?* (Fairfax, VA: Environmental and Toxicology International, n.d.).

22. R. I. Krieger et al., "Adult and Infant Abamectin Exposures Following Avert 310 and Pressurized Gel Crack and Crevice Treatment," *Bulletin of Environmental Contamination & Toxicology* 58 (1997): 681–687.

23. *Final Report: Nonoccupational Pesticide Exposure Study (NOPES)* (Washington, D.C.: Environmental Protection Agency, Office of Research and Development, 1990): 60.

24. "Home Garden Pesticide Survey Results," *Pesticide and Toxic Chemical News* 20 (10) (March 18, 1992): 33.

25. *Yardscaping: Minimizing Reliance on Pesticides by Example Using Demonstration, Outreach, and IPM Training* (Washington, D.C.: Pesticide Environmental Stewardship Program, n.d.).

26. Fred Whitford et al., *Pesticides and the Home, Lawn, and Garden* (West Lafayette, IN: Purdue Pesticide Programs, n.d.).

27. B. J. Bashin, "Bug Bomb Fallout," *Harrowsmith Journal* (May–June 1989): 45.

28. Z. R. Gard et al., "Why is Irritability, Anger, and Viciousness Increasing?" *Explore for the Professional* 6 (4) (1995): 39–45.

29. "Asthma, Allergies, Bronchitis, and Cough Linked to Chlordane Homes," *Bulletin of Environmental Contamination & Toxicology* 39 (1987): 903.

30. *Bulletin of Environmental Contamination & Toxicology* 27 (1981): 406.

31. J. Milton Clark, "Families in Apartment Complex Treated With Chlordane Show Increases in Sinusitis, Bronchitis, Migraines, Cough and Anemia," *Bulletin of Environmental Contamination & Toxicology* 39 (1987): 903.

32. Richard A. Cassidy and George M. Vaughn, "Breast Cancer Linked to Chlordane Exposure," *Breast Cancer Research and Treatment* 90 (2005): 55–64.

33. "Immune System Damage and Autoimmune Problems," *Archives of Environmental Health* 43 (5) (1988): 349–352.

34. Richard A. Cassidy et al., "Overweight—A Symptom of Chlordane Exposure," *Toxicology & Applied Pharmacology* 126 (1994): 326–337.

35. Rhonda J. Ferree, *Methyl Parathion Misuse: Questions and Answers* (Urbana-Champaign, IL: University of Illinois College of Agriculture, Consumer and Environmental Sciences, n.d.).

36. S. E. Lajeunesse et al., "A Homeowner Survey: Outdoor Pest Management Practices, Water Quality Awareness, and Preferred Learning Methods," *Journal of Natural Resources and Life Sciences Education* 26 (1) (1997): 43–48. D. C. Sklar et al., *In Colorado: A Survey of Woody Plant Nurseries and Homeowners, 1995–1996*, Technical Bulletin TB 97-2 (Fort Collins, CO: Colorado State University Cooperative Extension Service, 1997).

37. Kyle Cecil and George Czapar, "Urban Integrated Pest Management Training for Retail Store Employees," *Journal of Extension* 39 (1) (February 2001).

38. Natural Resources Defense Council, *Poisons on Pets* (Washington, D.C.: Natural Resources Defense Council, n.d.).

39. *"Top 10 Pesticide Blunders" Provide Cautionary Tales* (Sacramento, CA: California Department of Pesticide Regulations, April 18, 2000): 2.

40. "Florida Couple Wins Chlordane Contamination Case," *Pesticides and You* 18 (4) (1999): 8.

41. Patricia J. Wood, quoted in *Port Washington News* (NY), "Grassroots Environmental Education Partners with Long Island Counties," October 6, 2006.

42. *Healthy Lawn Care for Healthy Communities* (Montpelier, VT: Vermont Public Interest Research Group, n.d.).

43. B. Dubey et al., *Environmental Science and Technology* 38 (20) (2004): 5,400–5,404.

44. *Pesticide-Free Lawn Care* (San Francisco: Pesticide Action Network North America, n.d.).

45. *Issues We Work On* (Boston: Toxics Action Center, 2003).

46. *Environment and Human Health Inc. Is Releasing the Findings of Its Research Report on the Use of Lawn Care Pesticides* (North Haven, CT: EHHI, June 24, 2003).

47. *Carbofuran: A Special Review Technical Support Document* (Washington, D.C.: Environmental Protection Agency, Office of Pesticides and Toxic Substances, 1989).

48. J. R. Reigart and J. R. Roberts, *Recognition and Management of Pesticide Poisonings* (Washington, D.C.: Environmental Protection Agency, Office of Prevention, Pesticides, and Toxic Substances, 1999): 34–36.

49. D. M. Schreinemachers, "Birth Malformations and Other Adverse Perinatal Outcomes In Four U.S. Wheat-Producing States," *Environmental Health Perspectives* 111 (9) (July 2003): 1,259–1,264.

50. Francine Stephens, *Lawn and Garden Pesticides* (Los Angeles: Children's Health Environmental Coalition, n.d.).

51. "Lawn, Garden Pesticides Are Child Risk Indoors," *Albion Monitor*, February 18, 1997.

52. M. Nishioka et al., "Distribution of 2,4-D-dichlorophenoxy Acetic Acid in Floor Dust Throughout Homes Following Homeowner and Commercial Lawn Application: Quantitative Effects on Children, Pets, and Shoes," *Environmental Science & Technology* 33 (1999): 1,359–1,365.

53. K. G. Black and R. S. Fenske, "Dislodgeability of Chlorpyrifos and Fluorescent Tracer Residues on Turf: Comparison of Wipe and Foliar Wash Sampling Techniques," *Archives of Environmental Contamination Toxicology* 31 (4) (November 1996): 563–570.

54. Larry Schulze and Susan Schoneweis, *Pesticide Use and Safety* (Lincoln, NE: University of Nebraska Pesticide Education Resources, n.d.).

55. *Pesticide Packaging, Labeling, and Sales* (North Haven, CT: Environmental Human Health, Inc., n.d.).

56. Gary Korb, James Hovland, and Steven Bennett, *A Note of Caution on Fertilizer-Pesticide Combinations* (Madison, WI: University of Wisconsin Extension, 1999).

57. *Questions and Answers on Lawn Pesticides* (Washington, D.C.: Beyond Pesticides, n.d.).

58. "State Lawn Pesticide Notification Laws," *Pesticides and You* (Washington, D.C.: Beyond Pesticides/National Coalition Against the Misuse of Pesticides) 24 (2) (2004): 22.

59. K. H. Arne, "State Pesticide Regulatory Programs: Themes and Variations," *Occupational Medicine: State of the Art Reviews* 12 (2) (1997): 371–385.

60. Shawnee Hoover, "No Justice For Pesticide Victims," *Pesticides and You* 24 (1) (2004): 9–10.

61. D. B. Elgin, *Public Awareness Study: Summary Report* (Seattle, WA: The Water Quality Consortium, 1996).

62. *Organic Lawn Care* (Toronto: Toronto Environmental Allinace, n.d.).

63. Kay Lazar, "North Shore Activists are Allying with Towns to Limit Chemicals on Lawns, but Find Resistance from Old-School Weed Fighters," *The Toronto Globe*, May 1, 2005.

Six

Pesticides in the Air, Water, and Soil

Pesticides in the Air

> Imagine this scenario: It's two in the afternoon, and you're at home, just sitting down to enjoy a late lunch when the quiet atmosphere is broken by the drone of an approaching crop-dusting plane. But wait a minute ... what is that guy doing? Why, he's spraying practically right over your house! Worse yet, the wind is blowing toward you, and your children are playing in the yard.
>
> —Terry Shafer[1]

Introduction

Pesticides have been used for decades to control agricultural pests and to ensure an adequate quantity and quality of food for the nation. A pesticide's toxicity is responsible for its effectiveness in controlling pests, but the chemical may cause undesirable side effects when it travels from its intended location. Pesticide movement to the atmosphere depends upon complex interactions between the properties of the individual chemicals, the weather, the properties of the soil or plant tissue on which they are adsorbed, the way they are applied, and the management of the field or crop. Pesticides potentially can contaminate soil, water, and air.

The occurrence of pesticides in the atmosphere is an important national issue. Studies have documented that some pesticides found in the atmosphere and in water have resulted from agricultural applications. Dissipation and accumulation of pesticide residues can limit the efficacy of some pesticide materials. Pesticides and transformation products in the atmosphere can be major health concerns and cause plant damage far from their sites of application. For example, methyl bromide, a widely used soil fumigant, has been implicated in damage to the stratospheric ozone layer. More than half of applied materials may ultimately reach the atmosphere.

Today, pesticides have been detected in the atmosphere throughout the country, and a wide variety of pesticides are present in air, rain, snow, and fog. There is significant evidence that pesticides used in one part of the United States are carried through the atmosphere and deposited in other parts of the nation and beyond, sometimes in places where they are not even used. Even in the Arctic and Antarctic, pesticides are found in the air, snow, people, and animals. The extent of atmospheric pesticide contamination has not been adequately studied.

No one knows for sure just how many people nationwide have been sickened by pesticide drift. The federal government doesn't officially track such cases. But researchers at the EPA and the CDC have estimated that there are more than 5,000 serious poisonings a year from accidental drift.[2]

What is Pesticide Spray Drift?

Pesticide spray drift is the physical movement of a pesticide through air at the time of application or soon thereafter to any site other than that intended for application (often referred to as off-target sites). The EPA does not include the movement of pesticides to off-target sites caused by erosion, migration, volatility, or contaminated soil particles that are windblown after application, unless specifically addressed on pesticide product labels with respect to drift-control requirements.

How Does Spray Drift Occur?

When pesticide solutions are sprayed by ground-spray equipment or aircraft, the nozzles on the equipment produce droplets. Many of these droplets can be so small that they stay suspended in air and are carried by air currents until they contact a surface or drop to the ground. A number of factors influence drift including weather conditions, topography, the crop or area being sprayed, application equipment and methods, and decisions by the applicator.

Air Movement

Both horizontal and vertical air movement can affect drift. Unless it is calm, most pesticide applications are subjected to constant air movement. Indoors, heating and air conditioning systems move air and can move pesticides. Outdoors, unpredictable changes in air movement can cause spray drift at any time. Thus, wind direction and speed directly affect the direction, amount, and distance of drift.

The Impacts of Spray Drift

Off-target spray can affect human health and the environment. For example, spray drift can result in pesticide exposures to farmworkers, children playing outside, and wildlife and its habitat. Drift can also contaminate a home garden or another farmer's crops, causing illegal pesticide residues and/or plant damage. The proximity of individuals and sensitive sites to the pesticide application, the amounts of pesticide drift, and the toxicity of the pesticide are important factors in determining the potential impacts from drift. The drift of spray from pesticide applications can expose people, wildlife, and the environment to pesticide residues that can cause health and environmental effects as well as property damage.

Controlling drift is important for both commercial and private applicators. To be effective, the pesticide must be applied precisely on the target at the correct rate, volume, and pressure. Drift of herbicides can damage nearby crops, forests, or landscape plantings. Poorly timed applications can kill bees and other pollinators in the area. Beneficial parasites and predators that help control pests may also be killed. Indoor drift can also be a problem. Pest control operators must be aware that forced air heating systems and air conditioning units can transport sloppily applied pesticides.

Enforcement and Compliance of Laws

When individuals have complaints about off-target spray drift, they should report them to the state or tribal government agency (either agriculture or environmental protection) that is responsible for enforcing the proper use of pesticides for their state or tribe. These agencies are responsible for enforcing lawful use of pesticide products by investigating complaints and, when appropriate, issuing penalties for improper use. When necessary, the EPA will assist these agencies with investigations.[3]

Agricultural Spraying and Children's Exposure

Growing evidence exists that chronic exposure to low levels of organophosphate pesticides (OPs), widely used both in agriculture and residential settings, can cause adverse health effects in children. Despite these concerns, few studies have evaluated children's long-term exposure to OPs. A recent study examined year-long fluctuations in OP metabolite concentrations in a group of low-income children living in an agricultural community. The study found that regardless of their families' proximity to treated orchards or parental work exposure to pesticides, metabolite levels increased in children's urine during the spring and summer spraying months. Because OPs have a relatively short half-life in the body, levels declined (but were still detectable) in the fall and winter after agricultural spraying ended. Study results support the theory that children are continuously exposed to low levels of OPs in their diets, with episodes of higher exposures resulting from residential and agricultural pesticide use.[4]

Stricter Spray Drift Regulations

In mid-2002, the EPA announced its intention to restrict where and how farmers may spray their crops to prevent pesticides from poisoning farmworkers and residents of suburbs rapidly expanding into agricultural areas.

Pesticide companies fought the proposal, alleging that standards sought by the EPA were unwarranted and would cause 7 million acres of farmland to be taken out of production. Under rules to be spelled out on pesticide containers, the chemicals could not be allowed to drift on people, animals, homes, buildings, parks, wetlands, forests, pastures, or crops for which the spray was not intended. The labels would specify equipment sizes, the wind conditions under which spraying could take place, and the maximum distances from crops that spray could be released.

The EPA contended that the label rules would reduce the risks from pesticides without hurting farmers. But the agency had recently told the House Agriculture Committee that the standards proposed earlier would be revised before being made final.

States receive some 2,500 complaints annually about pesticide drift, and carry out enforcement actions on roughly 800 of them each year; the EPA estimates that there are probably many more unreported incidents. In 1999, 180 people in California's San Joaquin Valley were forced to evacuate when they were overcome by fumes from the spraying of a potato field. Months later, about thirty individuals still had respiratory problems, headaches, and dizziness.

In 2000, a herbicide that Bureau of Land Management employees were spraying on federal property in Idaho drifted onto nearby farmland and caused $100 million in damage to potato, wheat, and sugar beet crops.

Critics of the EPA's proposed rules questioned whether the problem was as serious as the EPA maintained, and argued that the standards did not take into account differences in topography and equipment.

In California, the number of spray-drift incidents in which at least one person was exposed to a pesticide dropped from ninety-four in 1995 to forty-one in 2000, the latest year for which figures were available. In comments filed with the EPA, the pesticide industry claimed the proposed rules would set a "zero-drift" policy for which pesticide use could not be maintained.

The deputy director of California's Department of Pesticide Regulation disagreed, stating that federal rules were necessary because states were not allowed to regulate pesticide labels. He said that having uniform standards nationwide "is better from our perspective as regulators but also for the industry."[5]

No Federal Action

The EPA has yet to act on controlling spray drift on a national scale. Despite the proposal to strengthen label provisions, no meaningful action followed. After being swamped by what the EPA called a "relatively huge response"—5,000 comments—and after extending the comments deadline twice, the agency decided to scrap the proposed change altogether and start over.

Environmentalists by and large supported the stronger label provisions, while arguing that even more needed to be accomplished. Not to be outdone, the pesticide industry claimed that these changes would be too expensive. Currently, U.S. pesticide law offers citizens no particular protection against spray drift. An EPA official stated, "The bottom line is that someone following the approved procedure for spraying a lawn or landscape with chemicals has the right to do it."

Nor do organic farmers marketing produce grown free of chemicals get any special protections, causing problems for some farmers who have lost their organic certification. One baby food company was outraged to learn that its products had been contaminated by pesticides through no fault of its own.

Homeowners who believe they have been exposed to pesticides must offer proof of harm, either with tests showing chemical residues on their property, or medical evidence. Many more homeowners are beginning to fight, but their cases are rarely publicized due to out-of-court settlements.[6]

Some methods for ranking agricultural pesticides by their potential hazard as air contaminants have been proposed based on use, volatility, toxicity, and other factors. Ultimately, rankings are used to determine exposure reduction or public health priorities. One of the initial uses of the ranking developed by the California Department of Pesticide Regulation, called the (pesticide) toxic air contaminant (TAC) ranking, is to direct air monitoring of agricultural pesticides in California. The California Air Resources Board conducts air monitoring in regional urban centers and in agricultural communities that are selected on the basis of area use of the monitored pesticides. For the monitored pesticides, an opportunity exists to calculate inhalation risk.[7]

Pesticides That Reach the Atmosphere

What happens to pesticides that are applied, and how much pesticide residue enters the atmosphere? This is not a simple question to answer. The fate of pesticides in the environment is dependent on many factors, such as their physical and chemical properties, the weather, and how, when, and where they were applied. Recent studies have shown that many pesticides readily evaporate into the atmosphere. Evaporation is a continuous process that occurs over weeks, months, and years, until all the pesticide molecules are degraded. Depending on the pesticide, 75 percent or more of an application can ultimately be lost through evaporation.

Annual deposition of selected pesticides by rain has been calculated in several areas of the country. The amount deposited generally accounts for less than 1 percent of the total applied. Although this seems like very little, it can represent many tons for some high-use pesticides. In addition, rain and snow are not the only way pesticides are deposited to the earth's surface. Deposition of vapors and particles also occurs, but there is an inadequate understanding of these dry deposition processes.[8]

Conclusions

The combined results from the local, region, and national monitoring studies indicate that a wide variety of pesticides are present in the atmosphere. Nearly every pesticide that has been investigated has been detected in air, rain, snow, or fog throughout the country at different times of the year. Also, there is ample evidence that some long-lived pesticides used in one area of the country migrate through the atmosphere and are deposited in other areas of the country, sometimes in areas where pesticides are not used. The atmosphere is an important part of the hydrologic cycle that can transport pesticides from their point of application and deposit them in unintended areas. Average annual concentrations of pesticides in air and rain are generally very low, although elevated concentrations may occur during periods of high use, usually in the spring and summer months. The environmental effects of long-term occurrences of low levels of pesticides in the atmosphere are not yet well understood.

Air has the ability to move particles over long distances. Most of the time this ability aids mankind. It causes rain, for example. Unfortunately for the pesticide applicator, wind also causes drift. Pesticides in the air are not controllable and may settle into waterways, homes, lawns, wooded areas, and so on. Drift must be avoided.[9]

Pesticide Names

Most pesticides that have been studied in the atmosphere have been detected, and many pesticides from several different chemical groups have been found at more than half the locations of samples nationwide. Results for different groups and individual pesticides reflect a range of influencing factors. Because of their widespread use during the 1960s and 1970s, and their resistance to environmental degradation, organochlorine insecticides have been detected in the atmosphere in every state where measurements were made. The most heavily used organochlorine insecticides during this time were toxaphene, DDT, and aldrin. Because of their reduced effectiveness and regulatory restrictions, their total use in agriculture declined steadily from 63 percent of insecticide use in 1966 to less than 5 percent in 1988. The most frequently detected organochlorine insecticides have been DDT, alpha-HCH, gamma-HCH (lindane), heptachlor, and dieldrin. Despite their widespread use, toxaphene and aldrin were detected less frequently than other organochlorine compounds, partially because of their chemical properties. Toxaphene is a complex mixture of more than 200 different compounds and is difficult to sample and analyze. Since toxaphene use was banned in 1982, the analytical "fingerprint" of environmental samples often differs considerably from analytical standards due to changes over time in its chemical nature. Also, the analytical limit of detection is much higher than for other organochlorine insecticides. On the other hand, aldrin in the environment degrades rapidly into dieldrin, which is more chemically stable. This is why dieldrin was detected more frequently than aldrin even though it was used in much lower quantities.

Organophosphorus insecticides also have been used heavily for decades and account for 65 percent of insecticide use today. Generally, they are not as long-lived in the environment as organochlorine insecticides, but nevertheless have been detected in the air and rain in many states, even though they are not often included as specific targets. The organophosphorus insecticides detected most often in air, rain, and fog were diazinon, methyl parathion, parathion, malathion, chlorpyrifos, and methidathion. Diazinon, methyl parathion, parathion, and malathion have been among the most widely used insecticides in each of the last three decades, although parathion, malathion, and diazinon use is declining.[10]

Aerial Spraying Developments

During aerial pesticide application, some of the applied material is lost to the atmosphere in the form of fine droplets moving off-target through the air stream by a process called spray drift. Spraying pesticides through spray nozzles produces a spectrum of droplets of differing diameters. The smallest droplets will remain airborne and become lost as spray drift. Larger droplets can be transported by the wind and deposited some distance outside the target area. As droplets are transported, their diameter decreases through evaporation. As they become smaller, they remain airborne longer and can be transported over regional, continental, or intercontinental distances.

Pesticides of moderate-to-high volatility sprayed above the soil surface form droplets that rapidly enter the gaseous phase and can be carried in the atmosphere. A portion of the pesticide that reaches the soil or plant surface also may evaporate over time and move into the atmosphere through a process of volatilization. Once in the atmosphere, a volatile pesticide can travel long distances. Loss during application through spray drift depends largely on the application method, properties of the formulation, and environmental conditions. Volatilization losses from soil or plants depend largely on soil and environmental conditions, chemical properties of the applied pesticide, and agricultural management after its application.

Once a pesticide is in the atmosphere, various atmospheric and chemical processes control its movement and transformation. Pesticides can break down in the atmosphere during photosynthesis or in reaction to other atmospheric constituents. Some processes are particularly important in determining the ultimate concentration and transport distance from the point of application, which affects the risk of contaminating sensitive ecosystems.

Source Reduction

Significant pesticide contamination is possible when pesticides are applied inappropriately or inefficiently or when accidentally spilled. Large quantities of applied pesticides may be lost from aerial spraying during windy conditions, and pesticides may drift onto adjacent fields or nearby ecosystems. Such conditions can cause significant atmospheric contamination.

Volatile pesticides are released into the atmosphere during and after application. Large amounts of pesticides may be released from areas of heavy agricultural activity for three to four days after application, causing increased pesticide concentrations in the entire region. Lower concentrations persist throughout the remainder of the year as the pesticide material is cycled within the plant-air-soil-water environment.

Potential impacts of pesticide loss to the atmosphere are 1) decline in air and water quality; 2) loss of beneficial insects and plants through off-site drift; 3) regional and long-range transport and degradation of soil, plant, and surface water quality; 4) accumulation and transfer of pesticide residues to sensitive wildlife and potential disruption of the food chain; and 5) degradation of the global atmosphere and loss of natural protective zones such as stratospheric ozone.

Many pesticides are volatile, and even those with low volatilities can be transported in the atmosphere as residues bound to dust particles or as aerosols. Both the active ingredient and formulation constituents can become air contaminants. Volatile components and residues bound to dusts may rise high into the atmosphere, travel long distances, and be deposited far from the point of origin through various deposition processes. Raindrops have been shown to have pesticide components.[11]

Air Pollution

When a pesticide does not turn up as a water contaminant, it is often because it has escaped to pollute the air instead. While no less troubling than water contamination, air contamination is significantly less well characterized and addressed. The EPA has stated that: "Off-target spray can affect human health and the environment.... There are thousands of complaints of off-target spray drift each year."[12] Though equipment modification and avoiding dangerous weather patterns can minimize drift, the EPA nonetheless notes that "some degree of drift of spray particles will occur from nearly all applications."[13] Air contamination from drift is thus an inevitable result of spraying and one to which regulatory programs are essentially unequipped to respond.

Vapor Movement

A pesticide that has vaporized (evaporated) can be carried from the treated area by air currents. Vapor movement, unlike spray or dust drift, is related to the chemical properties of the pesticide. Unlike the drift of sprays and dusts that can sometimes be seen during an application, vapor movement is not visible. Vapor movement can be caused by vapor leakage. Fumigants and other volatile materials exert pressure on the environment around them. Like air in a balloon, they are actively trying to escape. Keeping pesticide containers closed or sealed can stop vapor leakage. Fumigation sites must also be sealed properly to keep pesticides from leaking. Applying these materials with vapor-tight equipment is important. Some herbicides in particular can volatilize and move from a treated area, reducing control of the target weeds and increasing the likelihood that non-target plants will be injured. Pesticide vapors inside a dwelling can also cause injury, particularly if the occupants are sensitive.

Evaporation, while less obvious than drift, may actually be the largest single source of pesticides in the environment.[14] Unlike drift, evaporation is not limited to sprayed pesticides but occurs with liquid, powder, and granular pesticides as well. Once pesticides evaporate, they become part of the atmospheric water cycle. The U.S. Geological Survey states: "Nearly every pesticide that has been investigated has been detected in air, rain, snow, or fog throughout the country at different times of the year."[15] Like water contamination, air and precipitation monitoring show that local air detections of pesticides generally reflect pesticide use in the monitoring area, though some pesticides are carried far away from application sites and redeposited in areas where they have never been used. More-persistent pesticides, such as DDT, can travel in the upper atmosphere to the most remote locations on earth.[16]

If water quality standards offer little comfort that levels of pesticides are not causing health or ecological problems, the situation is less comforting still for air contamination. There are essentially no standards to provide a benchmark of exposure, and no regular programs to monitor pesticide levels in air, even if such standards existed.

Airborne Pesticide Contamination Threatens Human Health

A report entitled *Secondhand Pesticides: Airborne Pesticide Drift in California,* by Pesticide Action Network North America (PANNA), California Legal Rural Assistance Foundation (CLRAF), and Pesticide Education Center (PEL), in May 2003, revealed that several widely used pesticides were regularly found in air far from where they were applied at concentrations that significantly exceed levels deemed "safe" by regulatory agencies. The report demonstrated that current regulations ignore 80 to 95 percent of airborne movement of hazardous drift-prone pesticides, endangering the health of many hundreds of thousands of Californians.

The report revealed that pesticides are not only an immediate poisoning hazard for farmworkers and others directly exposed, but can adversely affect the health of people far from fields through the air they breathe. The report found that four of the six commonly used pesticides evaluated had concentrations in air at significant distances from fields that greatly exceeded the "acceptable" short-term "reference exposure levels" (RELs) for both children and adults. RELs are the concentrations of pesticides in air below which the EPA or California's Department of Pesticide Regulation (DPR) considers adverse health effects unlikely. Ongoing background exposure to pesticides in air in high pesticide use areas also poses considerable long-term health risks.

Near-field concentrations of chlorpyrifos and diazinon—both neurotoxic insecticides that the EPA is phasing out for home use because of the hazards they pose to children—exceeded the short-term child REL by 184 and thirty-nine times, respectively. For the highly acutely toxic fumigant metam sodium, concentrations more than 450 feet from the tested field exceeded the "acceptable" short-term child and adult REL by sixty times. Over the long term, lifetime cancer risks from exposure to average concentrations of the fumigant Telone in Kern County, California, measured up to fifty-six per million, far in excess of the cancer risk of one in one million that agencies generally consider the threshold for concern.

More than 90 percent of pesticides used in California are prone to drifting away from where they are applied. Of the 188 million pounds of pesticides used in 2000, 34 percent were highly toxic to humans. These are capable of triggering asthma and causing immediate poisoning and other respiratory illnesses, cancer, birth defects, sterility, neurotoxicity, and/or damage to the developing child.

Despite the health risks associated with widespread airborne pesticide drift, state and federal regulations ignore 80 to 95 percent of total movement of drift-prone pesticides by defining drift too narrowly: only spray drift that occurs during and immediately after an application is regulated. For 45 percent of pesticides applied in California, the concentrations of pesticides in air peak long after the application is complete—between eight and twenty-four hours after an application.

In addition to excluding most of drift problems, current drift regulations are ambiguous, and enforcement is difficult, weak, and largely ineffective.[17]

Legal Action to Enforce Toxic Air Contaminant Law

In June 2005, environmental health and community groups filed suit in Sacramento Superior Court to require that California's DPR uphold the Toxic Air Contaminant (TAC) law. The law, effective in 1984, requires the DPR to assess all pesticides as potential air contaminants and regulate them in order to protect public health.

Of the more than 900 pesticides registered in California, the DPR has completed the review process for only four in the past twenty years. The TAC statute is increasingly important because pesticides are a major component of air pollution in California's Central Valley and are one of the top three contributors to ozone pollution in the San Joaquin Valley, accounting for about 8 to 10 percent of the ozone-forming gases produced in the region.

High levels of ozone trigger asthma attacks and exacerbate other respiratory illnesses. In 2002, asthma rates in Fresno County were reported to be the highest in the state—ahead of even Los Angeles—and the third highest in the nation. Also, nearly one-third of pesticides used California are associated with serious chronic and acute health problems, such as cancer or nervous system maladies.

Pesticides are the largest source of toxic substances released into the environment in California. In 2002, pesticide use accounted for the release of 5.7 times more toxic materials to the environment than manufacturing, mining, or refining facilities as reported by the EPA's Toxic Release Inventory.

The lawsuit seeks to compel the DPR to comply with its duty under the TAC to assess all toxic pesticide air pollutants on a timely schedule, to take action to reduce the health impacts of these air pollutants, and to comply with the sections of the law requiring public transparency and input, including a review by an independent scientific review panel and substantive cooperation with the Air Resources Board and the Office of Environmental Health Hazard Assessment.[18]

Strict Liability for Spray Drift

In at least four states, Louisiana, Oklahoma, Oregon, and Washington, courts have labeled aerial application of pesticides an "ultra-hazardous" or "abnormally dangerous" activity, and have imposed strict liability for damage done without requiring proof of fault.

In 1957, Louisiana was the first state to impose strict liability for damages caused by the aerial application of pesticides. In the case of *Gotreaux v. Gary*, the defendant sprayed his rice crop with 2,4-D. This herbicide drifted onto the plaintiff's cotton and pea crops located more than three miles away and destroyed them. The court recognized the necessity of applying pesticides, but held that the plaintiff could not be unreasonably inconvenienced or denied the right to enjoy his property. The court summed up the effect of the application of strict liability: "negligence or fault, in these instances, is not a requisite to liability irrespective of the fact that the activities resulting in damages are conducted with . . . reasonable care and in accordance with modern and accepted methods."

In 1961, the Oregon Supreme Court in *Loe v. Lenhardt* imposed strict liability in an unintentional trespass suit, finding that there was no need to prove fault or negligence where the defendants were engaged in an "extra-hazardous" activity. In this case, the defendants were using a mixture of dinitro and diesel oil as an herbicide. The spray drifted, having, in the court's words, a "swift and drastic effect" on the plaintiff's crops.

The court, noting "the high degree of danger inherent in the spraying of agricultural chemicals from the aircraft," determined that strict liability should attach to the activity. The court stated the usual justification for the imposition of strict liability rather than a negligence standard: "element of fault, if it can be called that, lies in the deliberated choice by the defendant to inflict a degree of risk upon his neighbor, even though utmost care is observed in doing so."

The Washington State Supreme Court in 1977 imposed strict liability on crop spraying operations in *Langen v. Helicopters*. In this case the plaintiffs were organic farmers. The defendant's helicopter sprayed a neighboring farm with the pesticides Thiodan and Guthion. The plaintiffs sought damages for pesticides that drifted onto their crop of organically grown vegetables, rendering them worthless as certified organic produce.

The plaintiffs proceeded to destroy their crop, and filed a claim for full damages. In upholding a verdict in the plaintiffs' favor, the court applied the test for imposition of strict liability, and concluded that crop spraying was an abnormally dangerous activity, justifying the imposition of strict liability. In reaching this conclusion, the court stressed that there was no proof to suggest that it is possible to eliminate the risk of drift by the exercise of reasonable care. The court added that while aerial application was prevalent in the area, it was carried out by a relatively small number of people. In justifying its decision to impose strict liability, the court stated that those who perform useful but dangerous activities must be held accountable for any damages that result.[19]

Pesticide Drift Incidents Affecting Human Health

A family from White Hall, Maryland, has been exposed to numerous pesticide drift incidents. A neighboring farmer routinely sprays his fields with paraquat, 2,4-D, dicamba, and atrazine. On numerous occasions the farmer has failed to contain or manage pesticide spray drift, allowing it to move onto the family's property and into their home. Over the past eight years, their children have been exposed to these chemicals. They have complained of sore throats, headaches, and burning eyes during and after the applications. Informal conversations with the farmer failed to stop their exposure, so they turned to the Pesticide Regulation Department in the U.S. Department of Agriculture. To date, the Pesticide Regulation Department has failed to take any action to stop their exposure. Christmas tree seedlings and hardwoods have been burned and some completely destroyed by drift.

On the morning of November 8, 2000, children arriving at Mound Elementary School in Ventura, California, walked into a cloud of Lorsban, a pesticide containing the active ingredient chlorpyrifos, which had drifted from a neighboring lemon orchard onto school property. Two children were sent home because of symptoms of pesticide poisoning. Students and school staff complained of headaches, nausea, and dizziness associated with pesticide exposure.

A family from Los Angeles, California, was exposed to pesticides when aerial pesticide applications on a neighboring property drifted on their property. The mother was pregnant during two applications. Their daughter has had several developmental complications, including a partial cleft palate, constant movement of her eyes, and the need for open-heart surgery. Their windows were open at the time of the applications. Another child in daycare with the family's daughter has suffered similar health problems. There is no reasoning by their doctors and no family history of drugs or alcohol.

A schoolteacher in Sarasota, Florida, was concerned about pesticide drift from a citrus farm and a golf course into an elementary school with more than 500 students. She had to take medical leave from exposure to the pesticides that drifted into her school. About forty teachers complained of health effects but were afraid of speaking up about the exposure for fear of losing their jobs.

In August 2001, a woman in New Freedom, Pennsylvania, was driving by a home that was being sprayed with pesticides by the ChemLawn Company. The spraying was done on a regular basis at the home, which was by the road. The woman was already chemically sensitive from an earlier incident with two household chemicals. The drifting lawn pesticides caused her extreme dizziness and tremors. She has been chemically sensitive for eight years and cannot afford a lawyer.

In August 2001, the Centers for Disease Control and Prevention issued a report about more than 230 people who became sick after malathion was sprayed aerially during the Medfly Eradication Program, which began in Florida in 1998. A Florida couple, the Ruys, had to move after their home was contaminated by malathion that drifted from a citrus grove west of their property. A toxicologist, a medical toxicologist, and an immunologist recommended that the Ruys leave their home as a result of this contamination to avoid continued exposure to neurotoxic poisons found in their home.[20]

The Failure of Laws and Regulations

Most people who are sickened or whose property is contaminated or made unusable by other people's pesticide use have less protection and recourse under the law than someone whose property is defaced with paint (with the exception of plaintiffs in Louisiana, Oklahoma, Oregon, and Washington). In light of the illness, economic loss, and ecosystem disruption associated with pesticide drift, major changes must be made in the laws and regulations related to pesticide drift in order to protect public health and the environment.

The fact of the matter is that spray drift is poorly regulated by current state and federal laws and regulations. Post-application drift, which can occur for many days after an application, is barely regulated at all. It is not acknowledged by the EPA as a source of exposure except for the case of fumigant pesticides and mosquito fogging agents. Even then, protection measures to reduce exposures are in place only for a single fumigant, Telone.

Inadequate enforcement compounds the problem, making it easy for pesticide applicators to be careless with applications with little threat of punishment for violations. However, not all problems from drift are the result of illegal applications. Drift that occurs when applications are conducted in accordance with the label law also cause problems. Here is where changes in laws and regulations are most needed.

The EPA and other state agencies have the authority to regulate drift, with EPA policies setting the regulatory floor for states. States are authorized to create more stringent regulations if they wish. California has done so with respect to drift, and is somewhat ahead of most other states in this regard. However, even in California, regulations have not been successful in preventing acute poisoning or long-term exposures that exceed levels of concern. Thus, federal and state regulations covering pesticide drift largely fail to protect human health and the environment.[21]

Pesticides in Water

> High-quality water is more than the dream of the conservationalists, more than a political slogan: safe water, in the right quantity at the right place at the right time, is essential to health, recreation, and economic growth.
> —Edmund S. Muskie[22]

Background

Even though today's chemically intensive agriculture is partly responsible for providing abundant low-cost supplies of food and fiber, it has also created water-quality problems. When the chemical revolution first began there was little concern about environmental consequences. Scientific testing indicated that DDT and other agricultural chemicals were generally not harmful to humans if used as directed. By the mid-1960s, however, there was a growing awareness that some agricultural chemicals were damaging the environment, and possibly harming humans as well. Awareness that agricultural chemicals were not staying on the fields, but were being washed into streams and rivers and seeping into groundwater, came about with the development of sensitive chemical testing procedures. These procedures did not become available for organochlorine pesticides (such as DDT, DDE, aldrin, dieldrin, heptachlor, and chlordane) until the late 1960s. The DDT problem was known before that time largely because of bioaccumulation, resulting in detectable levels in animals high in the food chain. In addition, Rachel Carson's book *Silent Spring,* released in 1962, increased public awareness.

Today, pesticide levels in water are monitored routinely. Pesticide residues have been found in groundwater, surface water, and rainfall. The EPA began to emphasize groundwater monitoring for pesticides in 1979 following the discovery of DBCP and aldicarb in groundwater in several states. DBCP and aldicarb are dangerous because of their high toxicity. They are in the carbamate class and are nematocides use to control nematodes or parasitic worms which live in water. DBCP and aldicarb leach from agricultural soil into water. In 1985, thirty-eight states reported that agricultural activity was a known or suspected source of groundwater contamination within their borders.[23] Since then, several federal and state agencies have developed programs to sample water resources and test for the presence of agricultural chemicals. Results published to date have shown that chemicals used in agricultural production have been found in groundwater, sometimes at levels exceeding the EPA's drinking-water criteria.[24] Monitoring for pesticides in surface water was frequent in the 1960s and 1970s as studies were conducted that led to the banning of chlorinated hydrocarbon insecticides. Sampling in the 1980s and 1990s found that the four leading herbicides in use during that time—atrazine, metolachlor, alachlor, and cyanazine—were frequently detected in surface waters in agricultural regions.[25] The highest levels of contamination occurred after planting and during the early part of the growing season. Most of the pesticides commonly used presently and in the past have also been found in the atmosphere, including DDT, toxaphene, dieldrin, heptachlor, organophosphorous insecticides, triazine herbicides, alachlor, and metolachlor.[26] These airborne pesticides return to the ground with rainfall and further contribute to water contamination. A recent survey by the U.S. Geological Survey of pesticides in the nation's waters concluded that pesticides were common in surface and shallow groundwater in both urban and agricultural areas, but investigators were not able to determine if contamination is lessening or worsening.[27]

Pesticides in the Aquatic Environment

Although certain characteristics of pesticides are well known, their final characteristics after they reach a body of water are extremely difficult to estimate. It is necessary to conduct both field studies and laboratory testing of soil and water environments. Pesticide transport in soil and transfer to water together with probable impacts on water quality are determined by conducting detailed field surveys and water residue analyses.

There are several factors that influence a pesticide's potential to contaminate water:

- The ability of the pesticide to dissolve in water (solubility).

- Environmental factors, such as soil, weather, season, and distance to water sources.

- Application methods and other practices associated with the pesticide use.

Groundwater contamination is higher when there is no crop or a young crop. A large, actively growing crop has the ability to reduce pesticide concentration through a variety of mechanisms:

- Larger plants consume more water from the soil and therefore reduce the ability of a pesticide to migrate through the soil and enter streams or groundwater.

- Larger plants can collect precipitation that prevents pooling of water and run-off from the area.

- Root zones enrich the microbial community of the soil, which enhances biodegradation of the pesticide by bacteria.

The Safe Drinking Water Act sets standards for drinking water and mandates the EPA set Maximum Contamination Levels (MCLs) for a number of pesticides in public water supplies. Private water supplies are not monitored or regulated by this act and must be arranged privately. Since pesticides are most prevalent in agricultural areas where most residents obtain their drinking waster from private sources, it is up to the consumer or well owner to monitor contaminant levels.[28]

Importance of Surface Waters

Streams and reservoirs supply approximately 50 percent of the nation's drinking water, primarily in urban areas. Streams, reservoirs, lakes, and downstream estuaries are also vital aquatic ecosystems that provide important environmental and economic benefits. Surface waters are particularly vulnerable to pesticide contamination because runoff from most agricultural and urban areas, where pesticides are applied, drains into streams. Pesticides may also enter streams through wastewater discharges, atmospheric deposition, spills, and groundwater inflow. The uses and ecological significance of surface water, combined with its vulnerability to contamination, make it particularly important to understand the extent and significance of pesticides in this part of the hydrologic system.[29]

Significance to Water Quality

Under provisions of the Safe Drinking Water Act, the EPA established MCLs for concentrations of certain chemicals in drinking water. Of the currently used pesticides, only nine have established MCLs. Compliance with the Safe Drinking Water Act is determined by the annual average concentration of a specific contaminant in drinking water, based on quarterly sampling. While MCLs do not directly pertain to concentrations of pesticides in untreated surface waters, they provide benchmark values for comparisons, and they facilitate perspectives on the significance of the levels observed in surface waters.[30]

Pesticides in Wells

On a national scale, fewer than 2 percent of wells sampled in multistate studies were found to have pesticide concentrations above the established MCL. Due to repeated detection of various pesticides in U.S. wells, the EPA proposed a State Management Program that would control or ban those pesticides with the greatest potential to contaminate groundwater. Five pesticides were initially selected due to the frequency of their occurrence: alachlor, atrazine, cyanazine, metolachlor, and simazine. According to the EPA, they all have been detected in many states and have the potential to reach levels that exceed health standards. They are all associated with serious health effects, including cancer.

The five selected pesticides are herbicides that are used to control broadleaf weeds and grasses. The EPA estimates that between 200 and 250 million pounds of these herbicides are applied annually in the United States. Atrazine, simazine, and cyanazine are applied to agricultural land before and after planting. Alachlor and metolachlor are applied to soil prior to plant growth.[31]

Historical Study Efforts

Several large national and multistate studies of pesticides in rivers and streams were conducted between the late 1950s and the mid-1970s. These and most other studies during this period focused on organochlorine insecticides, such as DDT and dieldrin; a few phenoxy acid herbicides, such as 2,4-D; and organophosphorus insecticides, such as diazinon, all in use at the time. Use of organochlorine insecticides declined dramatically after the 1960s, while use of organophosphorus and carbamate insecticides increased. In addition, agricultural use of herbicides increased dramatically, from an estimated 84 million pounds in 1964 to more than 500 million pounds in 1992.

In response to changes in pesticide use, the number of different types of pesticides monitored in surface waters from the mid-1970s to the present has increased. The scale of monitoring studies has changed as well. The national and multistate studies conducted during the 1960s and 1970s have been largely replaced by state and local surveys, or by regional studies directed at specific river basins. Recent studies have been relatively short-term, and their geographic distribution is highly uneven. Iowa, California, Florida, and the Great Lakes region have been the most frequently studied areas. The most extensive regional studies have been conducted in the Mississippi

River Basin. Overall, there has been a steady increase in monitoring of pesticides in surface waters over the last several decades.

In recent years, the herbicides alachlor, atrazine, and simazine have frequently exceeded their MCLs in individual samples. A number of studies have shown that procedures commonly used at most water treatment plants have little effect on concentrations of these herbicides in water. Thus, drinking water derived from some surface water sources in the central United States likely contains concentrations of one or more of these compounds above the MCL for part of the year because of seasonal patterns. Annual mean concentrations, however, rarely exceed the MCL.

Our ability to assess the occurrence of pesticides in surface waters is limited by several factors. First, water quality criteria have not been established for most pesticides and pesticide transformation products, and existing criteria may be revised as more is learned about the toxicity of these compounds. Second, criteria are based on tests of individual pesticides and do not account for possible cumulative effects if several different pesticides are present. Finally, many pesticides and most transformation products have not been widely monitored in surface waters. These factors, and the lack of data about long-term trends, show significant gaps in our understanding of the extent and significance of pesticide contamination on surface waters. The results of this analysis indicate a need for long-term monitoring studies using a consistent study design and targeting more of the currently used pesticides and their transformation products.[32]

The Atrazine Danger

Atrazine's extensive use, persistence in soil, and mobility in water make it the most frequently detected pesticide in ground and surface water across the United States. As a result, drinking water is a common source of atrazine exposure, especially in agricultural regions. For example, testing has found atrazine in finished water from 97 percent of surface water–supplied drinking-water systems in Iowa. In addition, a recent survey of nearly 1,500 groundwater wells around the country detected atrazine in 23 percent of the samples, and found it to be among the most common pollutants detected.[33]

Health Effects of Atrazine

A growing body of toxicological and epidemiological evidence has raised concerns that chronic atrazine exposure may cause a variety of adverse human health effects. One epidemiological study found an association between maternal exposure to triazine herbicides in drinking water and increased incidence of developmental effects in newborns, including low birth weights. Reduced sperm counts, decreased sperm motility, and prostate inflammation have been observed in male laboratory rats exposed to atrazine. Endocrine disruption by atrazine and other triazine herbicides has also been reported in laboratory studies. Researchers have observed chromosomal damage in animal cell cultures exposed to atrazine in concentrations comparable to the federal drinking-water standard.[34]

Certain Populations at Risk

Pesticides such as atrazine pose the greatest risk to the developing fetus, infants, and children. Developing biological systems are more prone to chemical disruption, and immature metabolic systems are less able to detoxify pesticides. Children may be disproportionately exposed to atrazine because they drink more water than adults on a body-weight basis. Epidemiological and laboratory animal studies suggest that prenatal and nursing exposure to atrazine can cause abnormalities in the developing fetus and newborn offspring. These abnormalities can include intrauterine growth retardation, low birth weights, and higher rates of prostate inflammation in males.[35]

Regulation of Atrazine in Drinking Water

The EPA has set an MCL for atrazine in drinking water at 3 parts per billion (ppb). Nevertheless, groundwater monitoring has detected the herbicide at concentrations above the MCL in at least ten states. For public drinking-water supplies, regulators determine compliance by averaging quarterly measurements. In agricultural areas, however, this method can overlook spikes in atrazine levels that occur in spring and summer and result in short-term exposures to levels significantly above the standard.[36]

Factors that Affect the Fate of Pesticides in Water

Various processes affect the fate of pesticides following an application, disposal, or spill. The two basic processes are those that transfer chemicals or influence their movement, and those that degrade or break down chemicals. Water is involved in the primary transfer processes of runoff and leaching, but is much less involved in the degradation processes.

Runoff

Runoff occurs when water carries pesticides, either mixed in the water or bound to eroding soil, to off-target points. Rain carries pesticides from plant leaves to foliage near the ground and into the soil. The amount of pesticide runoff depends on the grade or slope of an area, the erodibility and texture of the soil, the soil moisture content, the amount and timing of irrigation or rainfall, and the properties of the pesticide. Agricultural runoff from crops is a concern because it may contain fertilizer and pesticides. Agricultural runoff enters our water sources by seeping through the soil to groundwater or entering streams as surface runoff.

Leaching

Some pesticides move through the soil and leach into groundwater. Several factors influence pesticide leaching. A pesticide that is easily dissolved in water moves with the water as it seeps through the soil. Soil structure and texture influence the rate and depth of pesticide leaching. Sandy and gravel soils have poor adsorption characteristics and allow water and pesticides to leach through quickly. A heavy clay soil does not allow for rapid leaching. Adsorption influences pesticide leaching because pesticides that are strongly attached to soil particles leach less. Leaching of pesticides from treated areas, mixing and rinsing sites, waste-disposal areas, and manufacturing facilities is a major groundwater concern.

Most of our activities change the quality of the underground water sources. Polluted water typically enters an aquifer in recharge water originating at the land's surface. Pollution can also be injected directly into an aquifer, for example, by back-siphoning directly into a well.[37]

Groundwater Protection

Groundwater is an important and still relatively untapped natural resource in the United States. Groundwater accounts for nearly all of our freshwater reserves, but only 20 percent of total water consumption and 50 percent of drinking-water use. However, groundwater contamination is rapidly becoming a serious and ubiquitous environmental concern. While groundwater contamination may result from mineralization or other natural processes, it is usually attributed to waste-disposal practices and industrial and agricultural activities.

Management of groundwater aquifers to satisfy drinking-water standards is a formidable task. It is difficult to monitor the movement of groundwater, and there are substantial time lags between emissions and detection of chemical residues. Once the aquifer is contaminated, residues may remain in the groundwater for long periods, and it is technically difficult and costly to treat the aquifer. At present, the best remedial actions are filtration at the wellhead, which can be extremely costly if contamination is widespread, or the use of substitute drinking-water sources.

Generally speaking, the best strategy for protecting groundwater supplies is to control source emissions. In most cases, the elimination of source emissions will eventually mitigate groundwater pollution. However, such drastic measures are often not necessary; it may be possible to meet groundwater quality standards without eliminating emissions. The problem policy makers face is establishing the relationship between on-site emissions and groundwater contaminant concentrations. Lacking this knowledge, state or local governments may lean toward a complete ban on a particular chemical. For example, when the pesticide aldicarb was detected in wells on Long Island, New York, it was subsequently banned from use by farmers. Assuming the drinking-water standard for aldicarb adequately protects consumers, it might be asked whether a "safe" application rate could have been established for this pesticide.[38]

Types of Groundwater Pollution

Under certain conditions, contaminants including soil nutrients, wastes, and chemicals can migrate to groundwater sources. Pesticides applied directly to a site may be moved downward with rain or irrigation water. This method of contamination is called non-point source pollution. When pesticides enter a well directly from spillage or back-siphonage and enter the groundwater directly, it is called point source pollution.

Twenty-two pesticides have been detected in U.S. wells, and up to eighty are estimated to have the potential for movement to groundwater under favorable conditions. More than half of the states have reported some pesticide contamination of groundwater.

Because agricultural runoff is a diffuse source of pollution, it is hard to control. It is the number-one category of non-point source pollution in rivers and lakes. Non-point source is most simply defined in contrast to point source pollution, which comes from a specific place such as a pipe or smokestack. Contaminant concentrations arising from industrial or other point sources can easily be measured at the "end of the pipe." Non-point source pollution is difficult to assess because the source is spread over a large area, as in agricultural or mining regions.

Because groundwater moves slowly, contaminants do not spread quickly. After pesticides reach groundwater, they may continue to break down, but at a much slower rate because light, heat, and oxygen are less available. Thus, they can remain underground in slow-moving columns for an indefinite period. When groundwater becomes contaminated, the polluted water may eventually appear in the surface water of streams, rivers, and lakes.

Although some organophosphorus compounds are highly toxic to humans, they generally break down rapidly in the environment and rarely have been found in groundwater. Another group that replaced the chlorinated hydrocarbons are carbamate pesticides, including aldicarb, carbofuran, and oxamyl. These compounds tend to be soluble in water and weakly adsorbed to soil. Consequently, if not degraded in the upper soil layers, they have a tendency to migrate to groundwater. The most significant occurrences of groundwater contamination have been by carbamate pesticides. Aldicarb has been detected in more than 2,000 wells on Long Island as well as in twelve other states, including Maine and New Jersey. As awareness has grown of the potential for pesticides to leach to groundwater, attention has focused on ways of changing registration and monitoring requirements to prevent such contamination from occurring in the future. Intensive studies have also been carried out in an attempt to determine what levels of pesticides are acceptable in water supplies.[39]

Unacceptable Risks

The Safe Drinking Water Act charges the EPA to protect public health by establishing allowable levels of contaminants in drinking water. These include a Maximum Contaminant Level Goal (MCLG) and an MCL for each regulated contaminant.

The former is based on health concerns, while the latter is based on economic feasibility and is the level that is enforceable by law. Currently, the EPA has set standards for seventy-eight different contaminants, twenty-nine of which are pesticides. That leaves another 271 pesticide ingredients for which no standards have been set.[40]

Even when MCLs have been established, the EPA does not base them on health factors but rather on economics. As a result, health problems do occur when people drink water contaminated with pesticides below the "acceptable" level. For instance, researchers studied Iowa communities served by a reservoir that was contaminated with 2 parts per billion of the herbicide atrazine, which is *below* the 3 ppb MCL set by the EPA. Overall, researchers found twice as many birth defects in communities that consume pesticide-contaminated water. Heart defects increased threefold, as did defects of the urinary and genital systems. Limb-reduction defects, arms or legs that do not develop to their normal length, increased almost sevenfold.[41]

Drinking-water standards also do not take into account health effects on the most vulnerable populations, such as children, the elderly, or people with immune system compromising diseases, such as AIDS. Moreover, drinking-water standards do not account for the effects of chemicals in combination when evidence demonstrates that mixtures of common pesticides—even at so-called low concentrations in drinking water—are implicated in damage to the nervous, immune, and hormone systems.[42] The point that should be stressed is that treatment of drinking water does not necessarily solve the problem because the technology is designed to remove only certain chemicals for which an MCL has been established.

Tracking Pesticide Use

In 1996, Congress amended the Safe Drinking Water Act to require the development of Source Water Assessment Plans (SWAPs). The amendments were meant to supplement the traditional approach of relying on water treatment with stronger efforts to protect drinking-water quality at its source. The heart of the amendment requires that states undertake more complete reviews of potential contaminants, and also requires that communities take action to prevent pollution.

To inventory the potential contaminants within a source-water area as required by SWAPs, communities need detailed information on which pesticides are used, and where, when, and in what amounts. The information must be site-specific enough so that water providers can target aggressive pollution prevention efforts to particular places that are most vulnerable, such as areas with high runoff or with soils prone to leaching. In turn, the pesticide-use data can help measure the effectiveness of efforts to prevent pollution.

Detailed information on pesticide use will also help water-monitoring efforts. Water providers need to know which pollutants to test for, such as pesticides that are used heavily in a particular source-water area. Testing only for chemicals that have established MCLs makes little sense in areas where other pesticides may be heavily used or where waterways are especially vulnerable to pollution from a particular pesticide.

Better information on pesticide use will also be extremely helpful in protecting private domestic wells and other water systems that are not subject to current drinking-water rules.

Prevention Is Key

The most complete story of trends in response to regulatory action and reduced pesticide use is the decline in organochlorine pesticide concentrations that followed reductions in use during the 1960s and bans on uses in the 1970s and 1980s. Concentrations of total DDT levels in fish, for example, decreased rapidly from the 1960s through the 1970s, and then more slowly during the 1980s and 1990s, as documented by data from the U.S. Fish and Wildlife Service and the National Water Quality Assessment (NAWQA) Program. Just as notable as the declines, however, is the finding that persistent organochlorine pesticide compounds still occur at levels greater than benchmarks for fish-eating wildlife in many urban and agricultural streams across the nation.

In contrast, NAWQA findings show that concentrations of relatively mobile and short-lived pesticides in stream water respond more rapidly to changes in use than the less mobile and more persistent organochlorine insecticides. For example, increases in acetochlor and decreases in alachlor are evident in streams in the Corn Belt, where acetochlor partially replaced alachlor for weed control in corn beginning in 1994. The changes in use were reflected quickly in stream concentrations, generally within one to two years. Similarly, concentrations of diazinon decreased significantly from 1998 to 2004 in five of seven urban and mixed-land-use streams sampled in the Northeast, consistent with the EPA-mandated phase-out of nonagricultural uses of diazinon that began in 2002.

Long-term and consistent data for assessing trends are essential for tracking water-quality responses to changes in pesticide use and management practices, for providing early warning of unanticipated problems, and for updating and improving models. Long-term monitoring is particularly important for assessing the occurrence of pesticides in groundwater and the occurrence of persistent compounds in streams because concentrations change slowly, sometimes taking decades to respond to changes in use.[43]

There is one final caveat. The best protection against drinking-water contamination by pesticides is *prevention*.

Pesticides in the Soil

Sometimes I couldn't stand how my eyes were watering and my throat hurt; I couldn't stand the gas. I would run outside the field to get some air. Now, I can't breathe well, and my vision is blurry.

—Jorge Fernandez[44]

Introduction

Pesticides are applied to the soil or to a crop. Many techniques can be used to apply a pesticide depending on the type of formulation, the timing of application, the pest to be controlled, and other soil management considerations. A pesticide can be injected into the soil as a fumigant, or into irrigation water, or it can be sprayed onto the soil surface. Crops can be sprayed, for example, with boom sprayers or tunnel sprayers or by aerial application, or they can be treated with specific pesticides. Seeds are sometimes treated with pesticides prior to planting. Pesticides also can be incorporated into other materials so that release of the active ingredient occurs over a longer period of time.

Soil fumigants are a special category of pesticides that are highly mobile in the soil-water-air environment. Because of environmental and health concerns, several fumigants, notably methyl bromide, have been banned during the last decade.

Pesticides also have the potential to damage important organisms in the soil. Research has shown that less than 1 percent of pesticides that are applied to crops actually reach their targets. The remainder can often, therefore, end up in soil.[45] Although this problem has not been widely researched, it is known that pesticides have the capacity to destroy earthworms, fungi, and bacteria. Soil organisms are vital to the proper functioning of agricultural systems. Most importantly, earthworms and microorganisms break down organic matter and make nitrogen and other nutrients accessible to plants. Some earthworm species are particularly vulnerable to the toxic effects of pesticides.[46]

Earthworms Are Beneficial

Although one acre of soil may hold up to 8 million earthworms, most people pay little attention to these productive and beneficial animals. They mostly go unnoticed from day to day, unless a heavy rain forces them to the surface of the soil, an angler needs some bait, or their casts (fecal matter) disrupt a game of golf.

Earthworms benefit the soil in many ways, primarily due to the physical and chemical effects of their casts and burrows. Earthworm casts, consisting of waste excreted after feeding, are composed mostly of soil mixed with digested plant residues. Casts modify soil structure by breaking larger structural units (plates and blocks) into finer, spherical granules. As plant material and soil passes through an earthworm's digestive system, its gizzard breaks down the particles into smaller fragments. These fragments, once excreted, are further decomposed by other worms and microorganisms. Earthworm casts can contribute up to 50 percent of the soil composition in some soils.

Many species of earthworms deposit their casts beneath the soil surface within their burrows, where casts contribute to the formation and development of the soil. Species that excavate permanent, vertical burrows, however, deposit their casts on the soil surface, where they play a greater role in soil development. In addition to benefiting soil structure, casts also provide nitrogen in a usable form for other organisms that decompose organic matter on the soil surface.[47]

Earthworms are generally found in the top twelve to eighteen inches of soil because this is where food is most abundant. Pesticides applied to control turf diseases or insect pests may severely affect earthworms. This can be avoided by accurately identifying and assessing problems and, if a treatment is necessary, selecting products that have the least detrimental effect. Products commonly used on turf areas vary greatly in their toxicity to earthworms. Some pesticides can cause severe and long-term reductions in earthworm numbers.[48]

Pesticide Breakdown (Degradation)

Pesticides degradation may cause special hazards in the environment. Some break down into toxic compounds before degrading further. Others may fail to break down because of unusual environmental conditions. All areas possess unique environmental conditions that influence the way chemicals degrade. Some of these environmental factors include soil texture, soil moisture, soil organic matter, air flow, temperature, rainfall, and the presence of plants and animals. For example, some herbicide rates must be adjusted according to the type of soil that they are applied to in order to be effective, or in some cases, to prevent crop injury.[49]

Pesticide Residues

Poison sprays and synthetic fertilizers often kill bacteria that are necessary to decompose organic wastes, which themselves create nutrient-rich soil. In addition, the poisonous sprays sink deeply into soil and groundwater, polluting streams, lakes, and aquatic life.

Large amounts of insecticides repeatedly sprayed on plants will eventually enter the soil, killing living matter there. These organisms include invisible bacteria, fungi, and algae, which break down plant residues to release minerals, carbon, and nitrogen. These organisms also include insects, which break down plant matter into new soil, and earthworms, which dig tunnels that aerate the soil. Pesticides can cause the soil to become useless for cultivation.[50]

Improper use of pesticides can seriously affect the soil's microbial community, with the same results as repeated cultivation. A soil depleted of its microscopic flora and fauna loses its ability to decompose organic matter and becomes less fertile. It will have a poorer structure and porosity, and be less hospitable to plants than a soil rich with life. A diversity of beneficial organisms can also help control organisms that harm certain plants.[51]

Soil Properties and Leaching Potential

The following soil properties affect pesticide leaching:

Organic Matter. When plant and animal material decomposes in or on the soil, a small amount of the material remains in the soil as very slowly degrading organic matter. This organic matter binds most pesticides very effectively. The more organic matter in the soil, the less likely a pesticide will leach through the soil.

Texture. The percentage of sand, silt, and clay in a soil determines its texture. Soil texture influences how fast water can move through it. The more sand there is in the soil, the easier it is for water and any contaminants (that is, pesticides) to move into groundwater.

Acidity (pH). Soil acidity, or pH, affects the chemical properties of many pesticides. As soil pH decreases, pesticides bind more to the clay in the soil and are filtered out of the percolating water. Also, pesticides are usually less soluble in water at lower pH values. Acidity is more important with some types of pesticides than others, but is less important overall than organic matter and texture.

Other geologic and environmental factors also affect pesticide leaching to groundwater. Depth from the soil surface to groundwater is very important. The closer the water is to the surface, the less chance there is for a pesticide to be filtered and broken down in the soil. Weather also plays an important role in many ways. Pesticides degrade faster in warm, moist soil than in cooler or drier soil. If heavy rainfall or irrigation occurs soon after a pesticide application, the percolating water can carry the pesticide deep into the soil, where it breaks down more slowly. Also, types of tillage practices can affect soil temperature, moisture, and water infiltration, all of which have an impact on pesticide degradation and leaching.[52]

Adsorption. A soil-adsorbed pesticide is less likely to volatilize, leach, or degrade. When pesticides are tightly bound to soil particles in highly adsorptive soil, they are less available for absorption by plants and microorganisms. However, soil-adsorbed pesticides can be lost by erosion. Understanding adsorption factors can reduce damage to sensitive plants, leaching to groundwater, and the presence of illegal residues in a food or in feed crops.

The behavior of pesticides under local environmental conditions is determined with special reference to soil movement and persistence. Studies have shown that the behavior of pesticides differs in various soils and under local conditions. It has been shown, for example, that pesticides tend to move more readily in certain soils, and extended half-lives of these compounds may be expected. Prolonged persistence of soils combined with high soil mobility is indicative of greater pollution potential of a pesticide.[53]

Pesticides Disrupt Agriculture

If pesticides interrupt or destroy the microbiotic activity in the soil, it becomes merely an anchor for plant material. In this "conventional" method of agriculture, which has been in use for only the past seventy-five years out of 10,000 years of recorded agriculture, plants can receive only air, water, and sunlight from their environment; everything else must be distributed to the plant by the farmer, often from inputs transported thousands of miles to reach the farm. Plants are commonly fed only the most basic elements of plant life and so are dependent on the farmer to fight all of nature's challenges: pests, diseases, and drought.

Synthetic pesticides not only kill soil microbes and leave toxic residues on food, they also threaten the health of farmworkers and disrupt natural ecosystems around the farm.[54]

Microbial Degradation

Some pesticides in soils are destroyed by microbial degradation. This occurs when microorganisms such as fungi and bacteria use a pesticide as food. Under the proper soil conditions, microbial degradation can be rapid and thorough. Conditions that favor microbial growth include warm temperatures, favorable pH levels, adequate soil moisture, aeration, and fertility. Adsorption also influences microbial degradation because adsorbed pesticides are less available to microorganisms, and therefore degrade slowly. Certain pesticides require higher application rates to compensate for pesticide losses through microbial degradation. In an extreme case of accelerated microbial degradation, pesticides that are normally effective for weeks suddenly become ineffective within days. In such a case, previous pesticide applications may have stimulated the buildup of certain microorganisms that were effective in rapidly degrading the pesticide.

Chemical Degradation

Chemical degradation is the breakdown of a pesticide by processes not involving a living organism. The adsorption of pesticides to the soil, soil pH levels, soil temperature, and soil moisture contribute to the rate and type of chemical reactions that occur. Many pesticides, especially OP insecticides, are susceptible to degradation by fluid decomposition in highly acidic soils or spray mixes. Because the products of chemical degradation are usually nontoxic or nonpesticidal, the amount of pesticide is reduced, as is its potency.

Photodegradation

Photodegradation is the breakdown of pesticides by sunlight. Pesticides applied to foliage, soil, or structures vary considerably in their stability when exposed to sunlight. Like other breakdown processes, photodegradation reduces the amount of chemicals present and lowers the level of pest control. Mechanical combination with soil during or after application, or by irrigation or rainfall following application, can reduce pesticide exposure to sunlight.[55]

Soil Fumigants

Fumigants are used on a wide range of annual and perennial crops, stored commodities, structures, and food-processing facilities to control insects, parasitic worms, plant pathogens, and weeds. Millions of pounds of fumigants are used to produce these crops each year. Because of their high application rates, the most widely used soil fumigants, methyl bromide, Telone, metam sodium, and chloropicrin, rank in the top twenty pesticides based on pounds applied per year. In 2001, metam sodium was the third most commonly used pesticide in the United States (57 to 62 million pounds), methyl bromide was the seventh most commonly used pesticide (20 to 25 million pounds), Telone was the eighth most commonly used pesticide (20 to 25 million pounds), and chloropicrin was the eighteenth most commonly used pesticide (5 to 9 million pounds).

Fumigants are needed in situations where the pest problem is so great that it would otherwise be technically or economically infeasible to grow a crop without the use of these chemicals. The largest uses of soil fumigants are in potatoes, tomatoes, tobacco, carrots, and strawberries to control plant pathogens, parasitic worms, and weeds.

Fumigants are formulated and applied in several ways. Granule formulations such as dazomet are applied to the soil surface and then watered into the soil or mechanically distributed. Liquid fumigants can be applied by directly injecting them into the soil or in some cases by injection into the irrigation system. Soil retention of the fumigant, and control of emissions, is improved by the use of tarpaulins or water seals.

In addition to soil uses, fumigants have two other important functions. First, fumigation prevents the introduction or spread of plant pests or noxious weeds into or within the United States. Under regulation, certain plants, fruits, vegetables, and other items must be treated before they may be moved into, or transported within, the country. Next, commodities, structures, and food-processing facilities are fumigated principally to control insects using the penetrating characteristics of gaseous methyl bromide.[56]

Soil Fumigant Hazards

Each soil fumigant pesticide is different, but all have the potential to move off-site following field applications. Surrounding air currents lead to the exposure of bystanders near treated areas and people far away from treated areas. Use of soil fumigants also results in exposure of those handling the pesticides or working in treated fields. Acute inhalation exposures of bystanders and workers appear to present the greatest concern.

Strawberries and Tomatoes

Strawberries and tomatoes are two crops with the most intensive use of soil fumigants because they are particularly vulnerable to several types of disease agents, insects, parasitic worms, and mites that conventional farmers largely control with fumigants. These crops also use the greatest amount of methyl bromide, an ozone-depleting chemical. In California alone in 2003, 3.7 million pounds of metam sodium were used on tomatoes. Yet other farmers have demonstrated that it is possible to farm strawberries and tomatoes in a cost-effective way without the use of these harmful chemicals.[57]

Persistence Factors

Persistence is an important part of pest control, since successful pest control requires knowledge of the persistence period to make subsequent applications. A persistent chemical is advantageous for long-term pest control because fewer applications are needed.

Commercial applicators must be familiar with the persistence of each pesticide that may be applied to soil, especially where adjacent areas may be affected or where treated soil is used to grow other plants. When different plants are rotated in the same soil, phytotoxicity, or unintentional pesticide damage to plants, can be a problem. This is because a pesticide used to control some pests on one plant may leave residues in the soil that will damage or kill another plant. Information on the persistence of a given pesticide can be found on its product label.

Phytotoxicity results in abnormal growth, leaf burn and drop, and discolored, curled, and spotted leaves. If phytotoxicity is severe, the plant may die. Phytotoxicity often resembles other problems such as insect damage, plant disease, and poor growing conditions such as insufficient moisture and improper fertilization. As with phytotoxicity, pesticide persistence beyond the intended period of pest control contributes to accidental plant injury.[58]

An Important Enzyme Discovery

Researchers at the University of North Carolina at Chapel Hill have found that an enzyme (CaaD) inside a bacterium that grows in the soil of potato fields can—in a split second—break down residues of a common powerful pesticide, 1,3 dichloropropene, used for killing worms on potatoes.

Although it is expensive for farmers, if this particular enzyme were not in the soil, it would take 10,000 years for just half of the widely used pesticide to decompose. Also, the chemical would remain in the soil of the potato fields where it is now used in huge amounts, contaminating groundwater and posing a threat to human and animal health.

An unusual collaboration between an undergraduate student and a distinguished biochemistry professor at the university resulted in this important discovery. Dr. Richard V. Wolfenden, Alumni Distinguished professor of biochemistry and biophysics at the University of North Carolina School of Medicine, stated: "The half-life of the pesticide is longer, by several orders of magnitude, than the half-lives of other known environmental pollutants in water. The half lives of atrazine, aziridine, paraoxon, and 1,2-dichloroethane, for example, are five months, fifty-two hours, thirteen months, and seventy-two years, respectively." In contrast, he noted, the half-life of the potato pesticide residue, chloroacrylate, is 10,000 years, the same as the half-life of plutonium-239, the hazardous isotope produced in nuclear power plants.[59]

The Present State of Affairs

Plants draw minerals and other nourishment from the soil up through their roots, and these minerals are deposited in the body, fruit, and seeds of the plant. Plants then use these nutrients to manufacture the vitamins and nutrients humans consume from plants. Historically, farmers carefully tended and fed the soil through a variety of natural methods to keep it vital and healthy and to replenish the nutrients used by each crop. This rich soil in turn produced vital, healthy, nutrient-rich plants. With the advent of pesticides, natural soil feeding and tending declined markedly, and, as a result, there is little nourishment left in soil for plants to absorb. This has led to a marked reduction in the nutrient content of our food.

Pesticides kill not only insects that are regarded as a nuisance, but also beneficial insects that eat crop-destroying insects, as well as insect-eating birds and fish. Additionally, pesticides destroy the beneficial bacteria, insects, and worms that live in soil. These important organisms keep the soil vibrant and alive and enable strong, healthy plants to grow and flourish. Compromised soil leads to weak, unhealthy plants that have poor resistances to insects and diseases. The 1992 Earth Summit reported that the United States now has the worst soil on the planet. Eighty-five percent of our soil has been depleted to the point that it can no longer nourish healthy plants, and this has led to the problem of vanishing nutrients.[60]

Insects are extremely adaptable. Unlike humans, insects have brief life spans and reproduce prolifically, allowing them to quickly accommodate changes in their external environment though genetic mutations. As a result, insects have successfully survived numerous catastrophic planetary changes that wiped out other species. Our actions have also given them a competitive edge.

Notes

1. Terry Shafer, "If You Are Sprayed with Pesticides," *Mother Earth News,* May–June 1983.

2. David Brancaccio, "New Suburbs and Pesticides," PBS's *NOW,* August 19, 2005.

3. *Spray Drift of Pesticides* (Washington, D.C.: Environmental Protection Agency, Office of Pesticide Programs, December 1999).

4. D. Koch et al., "Temporal Association of Children's Pesticide Exposure and Agricultural Spraying: Report of a Longitudinal Biological Monitoring Study," *Environmental Health Perspectives* 110 (8) (2002): 829–833.

5. Philip Brasher, "EPA Seeks to Crack Down on Spray Drift from Farm Pesticides," Associated Press, April 6, 2002.

6. Francesca Lyman, "Does Pesticide Drift Pose Risks for Home Gardens?" MSNBC (May 8, 2002).

7. Robert Gunier et al., "Community Exposures to Airborne Agricultural Pesticides in California: Ranking of Inhalation Risks," *Environmental Health Perspectives* 110 (12) (December 2002): 1,178–1,184.

8. *Pesticides in the Atmosphere* (Washington, D.C.: U.S. Geological Survey, Fact Sheet FS-152-95, 1995): 152–195.

9. Ibid.

10. Ibid.

11. *Component V: Pesticides and Other Synthetic Organic Chemicals* (Washington, D.C.: U.S. Department of Agriculture, Agricultural Research Service, n.d.)

12. *For Your Information: Spray Drift of Pesticides* (Washington, D.C.: Environmental Protection Agency, 1999).

13. Ibid.

14. R. Huskes and K. Levsen, "Pesticides in Rain," *Chemosphere* 35 (12) (1987): 3,013–3,024.

15. *Pesticides in the Atmosphere*. Op. cit.

16. E. Atlas and C. S. Giam, "Global Transport of Organic Pollutants: Ambient Concentrations in the Remote Marine Atmosphere," *Science* 21 (9) (1981): 163–165.

17. "Airborne Pesticide Pollution Regularly Exceeds 'Acceptable Level', " PANNA press release (San Francisco: Pesticide Action Network North America, May 7, 2003).

18. "State's Failure to Prevent Pesticide Air Pollution Linked to Health Problems for California Residents," PANNA and Californians for Pesticide Reform press release (Sacramento, CA: Pesticide Action Network North America and Californians for Pesticide Reform, January 19, 2005).

19. *Laws Governing the Use and Impact of Agricultural Chemicals: An Overview* (Gainesville, FL: University of Florida, University of Florida Cooperative Extension Service Bulletin 311, 1995).

20. U.S. EPA PR Notice 2001-X, *Spray and Dust Drift Label Statements for Pesticides* (Washington, D.C.: Letter to Environmental Protection Agency, Office of Pesticide Programs, March 27, 2002).

21. *Secondhand Pesticides: Airborne Pesticide Drift in California* (Sacramento, CA: Californians for Pesticide Reform, May 7, 2003).

22. Senator Edmund S. Muskie, speech, March 1, 1966.

23. *America's Clean Water: The States' Nonpoint Source Assessment 1985* (Washington, D.C.: Association of State and Interstate Water Pollution Control Administrators, 1985).

24. *National Pesticide Survey Project Summary* (Washington, D.C.: Environmental Protection Agency, Fall 1990).

25. *Atrazine in Surface Waters: A Report of the Atrazine Task Group to the Working Group on Water Quality* (Washington, D.C.: U.S. Department of Agriculture, April 1992).

26. *Pesticides in the Atmosphere*. Op. cit.

27. *The Quality of Our Nation's Waters—Nutrients and Pesticides* (Washington, D.C.: U.S. Geological Survey Circular 1225, 1985).

28. D. I. Gustafson, *Pesticides in Drinking Water* (New York: Van Nostrand Reinhold, 1993).

29. *Pesticides in Surface Water* (Washington, D.C.: U.S. Geological Survey Fact Sheet FS-039-97, 1997).

30. Ibid.

31. J. E. Barbash and E. A. Resek, *Pesticides in Groundwater, Volume 2 of the Series Pesticides in the Hydrologic Environment* (Chelsea, MI: Ann Arbor Press, Inc., 1996).

32. *Pesticides in Surface Water.* Op. cit.

33. *Revised Preliminary Human Health Risk Assessment for Atrazine* (Washington, D.C.: Environmental Protection Agency, Office of Pesticide Programs, 2002).

34. R. J. Cooper et al., "Atrazine Disrupts the Hypothalamic Control of Pituitary-Ovarian Function," *Technological Sciences* 53 (2000): 297–307.

35. National Research Council, *Pesticides in the Diets of Infants and Children* (Washington, D.C.: National Academy Press, 1993).

36. *National Primary Drinking Water Regulations, Technical Fact Sheet on Atrazine* (Washington, D.C.: Environmental Protection Agency, Office of Ground Water and Drinking Water, 1998).

37. Jay Gan, "Pesticide Groundwater Quality," *Pesticide Wise* (Winter 2002).

38. G. D. Anderson, J. J. Opaluch, and W. M. Sullivan, "Nonpoint Agricultural Pollution: Pesticide Contamination of Groundwater Supplies," *American Journal of Agricultural Economics* 67 (1985): 1,238–1,243.

39. *Nonpoint Source Pollution: The Nation's Largest Water Quality Problem, Nonpoint Source Pointer No. 1* (Washington, D.C.: Environmental Protection Agency, Office of Water, 2004).

40. *Current Drinking Water Standards* (Washington, D.C.: Environmental Protection Agency, Office of Ground Water and Drinking Water, 1999).

41. R. Munger et al., "Birth Defects and Pesticide Contaminated Water Supplies in Iowa," *American Journal of Epidemiology* 136 (1992): 959.

42. W. P. Porter et al., "Groundwater Pesticides: Interactive Effects of Low Concentrations of Carbamates, Aldicarb, and Methomyl and the Triazine Mertibuzin on Thyroxine and Somatropin Levels in White Rats," *Journal of Toxicology and Environmental Health* 40 (1993): 15–34.

43. *Pesticides in the Nation's Streams and Ground Water, 1992–2001* (Washington, D.C.: U.S. Geological Survey, Department of the Interior, 2004).

44. Jorge Fernandez, farmworker poisoned by methyl bromide, Salinas, CA.

45. *Beneath the Bottom Line: Agricultural Approaches to Reduce Agrichemical Contamination of Groundwater* (Washington, D.C.: U.S. Congress, Office of Technology Assessment, November 1990): 104.

46. Robert L. Bugg, "Earthworm Update," *Sustainable Agriculture Technical Reviews* 6 (3) (Davis, CA: University of California, Sustainable Agriculture Research and Education Program, Summer 1994): 3.

47. Karen Delahaut and C. F. Koval, *Turf: Earthworms—Beneficials or Pests?* (Madison, WI: University of Wisconsin, n.d.)

48. D. A. Potter et al., "Toxicity of Pesticides to Earthworms and Effects on Thatch Degradation in Kentucky Blue Grass Turf," *Journal of Economic Entomology* 83 (6) (1990): 2,352–2,360.

49. Frederick M. Fishel, *Pesticide Residues* (Gainesville, FL: University of Florida, Institute of Food & Agricultural Science, September 2005).

50. *Pesticides and Groundwater Contamination Bulletin* (Columbus, OH: The Ohio State University, Bulletin 820, n.d.).

51. Grace Gershuny and Joseph Smille, *The Soul of Soil: A Guide to Ecological Soil Management* (Davis, CA: agAccess, 1995).

52. D. W. Goss, "Screening Procedure for Soils and Pesticides for Potential Water Quality Impact," *Weed Technology* 6 (33) (1992): 707–708.

53. P. J. Maret et al., *The Safe and Effective Use of Pesticides* (Berkeley, CA: Statewide Integrated Pest Management Project, Division of Agriculture and Natural Resources, Publication 3324, 1988).

54. *Why Certified Organic Food Is Better Food* (Unity, ME: Maine Organic Farmers and Gardeners Association, n.d.)

55. Patricia Muir, *Why Be Concerned About Pesticides?* (Corvallis, OR: Oregon State University, October 18, 2004).

56. *Overview of the Use and Usage of Soil Fumigants* (Washington, D.C.: Environmental Protection Agency, Office of Prevention, Pesticides and Toxic Substances, June 20, 2005): 4.

57. Lucius McSherry and Katherine Mills, "Strawberry and Tomato Farming without Fumigants and Other Toxic Pesticides," *Global Pesticide Campaigner* 15 (2) (August 2005).

58. Karen Delahaut, *Pesticide Phytotoxicity* (Madison, WI: University of Wisconsin, n.d.).

59. David Williamson, "New Discovery: If It Weren't for this Enzyme, Decomposing Pesticide Would Take Millennia," *UNC News Services,* October 24, 2005.

60. Lisa Lefferts, "Pass the Pesticides," *Nutrition Action Newsletter* (April 1989): 1–2.

Seven

International Trade in Pesticides

Some pesticides are just too dangerous to be used safely under the conditions of use in many developing countries.

—Professor Hermann Waibel[1]

Background

The expansion of free trade among nations has focused attention on issues related to the export and import of pesticides, especially when the pesticide in question is not registered for use within the United States. The export and import of pesticides is addressed in Section 17 of FIFRA, and labels of pesticide products exported but not registered for use within the United States must include the statement "Not Registered for Use in the United States of America." When food products that have been treated with such exported products are imported back into the United States, the cycle is often termed the "circle of poison." In practice, the frequency of imported food products that have been treated in conflict with EPA registrations or FDA tolerances is relatively low. The major problem is differences in standards established by different countries, since the United States has one system of standards, while most other countries follow an international standard established by the Codex Alimentarius Commission (a joint commission under the Food Agriculture Organization [FAO] and World Health Organization [WHO], both agencies of the United Nations). Thus, the primary challenge is coordinating standards applicable to pesticide residues allowed in treated food products.

The fact of the matter is that the United States has shown minimal interest in advancing efforts to mitigate the harm caused by pesticides traded in the global marketplace. In 1993, the EPA issued the U.S. pesticide export policy (hereafter referred to as the 1993 policy), which is essentially a plan of noninterference: it allows U.S. pesticide producers to sell their products abroad with few procedural restrictions. Presently, the 1993 policy gives special treatment to "unregistered pesticides." This class includes both those pesticides that the EPA has banned from domestic use and those which have never been submitted for EPA evaluation. The 1993 policy allows U.S. companies to produce and export both types of unregistered pesticides to any country so long as they are labeled "unregistered" and the importer is notified of this classification.

Developing countries that import unregistered pesticides suffer immeasurable damages from these products. Further, it is commonly understood that developing countries allow the importation of unregistered pesticides only because they, unlike the developed countries that produce those pesticides, lack a regulatory infrastructure that would enable them to make sound risk/benefit analyses regarding the use of such products. Some observers claim it is wrong for developed countries to continue to "push" their unregistered pesticides on their less-sophisticated and more-desperate neighbors. The 1993 policy tacitly endorses this "pushing."[2]

The Scope of Pesticide Exports

The United States is a major exporter of pesticides. Nearly 3.2 billion pounds of pesticide products were exported from U.S. ports between 1997 and 2000, according to an analysis of U.S. Customs records. This average rate of almost 2.2 million pounds per day—or forty-five tons per hour—represents a 15 percent increase over the average rate of 936 tons per day documented for the years 1992–1996. Between 1997 and 2000, the United States exported nearly 65 million pounds of pesticides that are banned or severely restricted domestically.

Ominously, the United States exported nearly 1.1 billion pounds of pesticides that have been identified as known or suspected carcinogens, an average rate of almost sixteen tons per hour.

The data were gathered using commercial transcriptions of U.S. Customs records of shipments from U.S. ports. Although this is the most comprehensive source of export information available in the public record, it remains only a partial source of production and trade information since many details are protected as trade secrets.[3]

International Regulation

The problems of pesticide use in developing countries are widely acknowledged by governments and international agencies. Since the 1980s a number of initiatives have been developed to reduce the risks. These include:

International Code of Conduct on the Distribution and Use of Pesticides

The Code of Conduct is addressed to importing and exporting governments, and to industry and public interest groups. It was negotiated by governments through the FAO of the United Nations and ratified in 1985. A 1994 survey carried out on the effectiveness of the code found that health issues caused by pesticides had not been reduced, and that environmental problems appeared to have worsened, though this may also reflect increased consciousness of environmental hazards related to pesticides.

The Rotterdam Convention on Prior Informed Consent

In industrialized countries, governments are able to test and assess pesticide hazards and risks and may ban or restrict those suspected of causing unacceptable health or environmental harm. Developing countries need early-warning systems to alert them to these actions. The 1992 Earth Summit recommended that the voluntary Prior Informed Consent (PIC) clause in the FAO Code become an international convention; it was ratified in September 1998. PIC covers pesticides that are banned or severely restricted for health or environmental reasons. Once included, governments must indicate whether they prohibit or consent to import. If they do not respond, it will be assumed that importation is not permitted.

PIC remained a voluntary procedure until Armenia ratified the convention on November 26, 2003, completing the fifty-nation ratification requirement, bringing the PIC treaty into force on February 24, 2004, ninety days after ratification. This means that the PIC treaty is a legally binding law.

Pesticides banned or severely restricted due to their health or environmental impacts can be included in the PIC procedure by the participating governments. Besides this, severely hazardous and acutely toxic pesticides (WHO Class 1a) that are a threat under the conditions of use in developing countries or countries with economies in transition may also be included.

The ratification of the PIC treaty is a significant event toward better protection of human health and the environment. This is an indication of a move toward a more precautionary approach in managing hazardous chemicals. The new legal strength of the treaty should concern all those involved in the production, distribution, and use of hazardous chemicals, especially pesticides. Although the treaty addresses chemicals in general, out of the thirty-two enlisted chemicals, twenty-seven are pesticides.[4]

Persistent Organic Pollutants

Since chemicals are so highly persistent in the environment that they cross national boundaries, moving from tropical regions to build up in the northern temperate areas, they also build up in the food chain and the fatty tissue of animals, including mammals. Nine of the twelve Persistent Organic Pollutants (POPs) identified so far are pesticides. These include DDT, still in use against mosquitoes, though most other POP pesticides are no longer available. Governments are negotiating a convention to phase out the production and use of POPs.[5]

Incomplete Records

There are indications that trade agreements are creating pressure for developing countries to increase their use of outdated, inexpensive, and hazardous products. The ongoing liberalization of trade has caused an influx of hazardous pesticides into developing countries. Trans-shipments have made it very difficult to know exactly where pesticides trade originates. For example, some of the products from the United States may come through European countries.

The U.S. government does not maintain complete records of pesticide shipments, and there are many data gaps. For instance, between 1992 and 1996, more than 2 billion pounds of pesticides left U.S. ports with their specific chemical names omitted from publicly accessible shipping records. The practice of manufacturers hiding their identity on exported products to prevent competitors from receiving confidential marketing information is legal in the United States, but creates an obstacle for developing countries and special-interest groups trying to expose the risks posed by the careless use of pesticides. Masking the identity of pesticides in customs records is a common illegal practice.[6]

Consequences of Lack of Awareness

Many pesticides used in developing countries are banned or severely restricted in the industrialized world because of safety concerns. These concerns are generally not shared by pesticide users in the developing world, due to a widespread lack of awareness of the hazards of pesticide exposure. Pesticide labels leave much to be desired. They are commonly unclear, written in a foreign language, lack clear health warnings, or are difficult or impossible to comprehend, especially by farmers, many of whom have poor English literacy. Recommended safety measures are often not employed. The use of protective masks, gloves, and boots is often impractical or simply unaffordable. Pesticides are frequently mixed, stored, or disposed of in a dangerous fashion, and are often applied too frequently or at too high a concentration.

Foreign farmers are naïve and will assume that any chemical coming from the United States must be relatively benign, despite the fact that most pesticides registered in this country would be too toxic to license if the EPA did not restrict their use.[7]

Small Pesticide Vendors

Small vendors are often ignorant of pesticide dangers because they have little workplace training. They fail to protect both themselves and their staff. They rarely provide workers with protective equipment, and often leftover pesticides are simply spilled onto the streets or into their backyards. Some pesticide distributors have credit systems that they offer to farmers to distribute and promote sales of pesticides. Competing pesticide distributors have their own extension agents whose sole purpose is to sell pesticides. Since the income of the pesticide salespersons depends on quantities they sell per season or day, each seller strives to be the top salesperson. In the process, pesticides are often misused and accumulate in the environment. Pesticide containers prove to be equally as dangerous as the pesticides themselves. In many countries, farmers are advised by pesticide distributing agents to bury containers in their backyards. Unfortunately, most of them end up for domestic use either as water containers or for food storage.[8]

Pesticide Exposure Numbers

As is obvious, pesticide poisoning in developing countries is quite frequent. In 2004, the UN and WHO estimated that 1 million to 5 million cases of pesticide poisonings occur each year, resulting in several thousand fatalities, including children. Most of the poisonings take place in rural areas, where safeguards are typically inadequate or nonexistent. Although developing countries use 25 percent of the world's pesticides production, they experience 99 percent of the resultant deaths.

Children are at higher risk because they are more susceptible than adults to pesticide exposure. Children's behavior, playing, and ignorance of risks result in greater potential for exposure. Malnutrition and dehydration also increase their sensitivity to pesticides. The levels of pesticides that persons in industrialized nations are exposed to are considerably lower than the levels in those who grow food for them, and their families, are exposed to.[9]

"Safe Use" versus "Safer Use"

The "safe pesticide use" slogan, particularly as promoted by pesticide manufacturers, has been the feature of a common approach toward mitigating the health problems caused by pesticides. Invariably, however, "safe use" neglects to describe the many alternatives available to farmers, choices including pesticides but not limited to them. Rather, the approach focuses only on pesticide-related matters, such as pesticide selection, new and correct application techniques and methodologies, registration issues, and the use of personal protective equipment such as masks and clothing. Consequently, many pesticide users, whatever their education, view pesticides as the crop protection method of choice, a "silver bullet," so to speak, when, in fact, there may be many less-toxic choices that are never even considered.

In reality, overwhelming evidence from the developing world demonstrates that "safe pesticide use" programs are not wholly successful. This is true whether or not studies measure utilization of safety gear and practices or rates of exposure. Of particular concern is the paradoxical fact that the use of protective equipment often increases personal exposure to pesticides. Poor user habits due to lack of water, soap, or initiative enable pesticides to accumulate in protective clothing and masks. Users are then subject to more exposure and higher doses with each "safety equipment" use.

Most farmers do not use safety equipment or gloves when applying pesticides. Even if they do, toxic residues remain from previous sprayings because they have too little water for cleaning and lack cautionary training. They also usually do not know how to adjust the sprayers properly in order to apply the correct amount of pesticide. As a result, too much is applied, wasting the pesticide and heightening the danger to people and the environment.

On the other hand, the new focus on "safer pesticide use" promotes the principle that all options for pest management should be considered, tested, and integrated into strategies for sustainable and environmentally sound crop production. A longer-term view of production and pest management is favored over the short-term reactive view. IPM provides farmers with the most choices and recognizes that pesticides may be effective in short-term or emergency situations, but maintains that they should be a tool of last resort due to their many unintended effects on health and the environment. The challenge is to maximize their effectiveness when they do have to be used, while reducing the risks of damage to human and environmental health as much as possible.[10]

Misuse of Pesticides

The FAO reported the widespread misuse of pesticides in the Third World. These unsafe practices were primarily associated with applicator exposure and the result of poor application equipment, but also suggested that environmental damage, pesticide waste, and excessive pesticide residues on foods were serious concerns. The FAO stressed the need for minimum standards for the safe and efficient application of agricultural chemicals and indicated that improvements in equipment quality and better training for farmers/applicators would dramatically improve the situation.

Farmers and applicators generally do not have sufficient knowledge about the pesticides or application techniques to use chemical technology safely. Many farmers believe that high spray-carrier volumes, high pressures, and high application doses are the correct way to use pesticides. Application equipment is not maintained, nozzles are not replaced, and hoses leak, resulting in environmental and applicator contamination.

Fifty percent of the pesticides applied in Pakistan were wasted due to poor application equipment and inappropriate use. India has levels of pesticide residues in food crops much higher than the world average, thus indicating incorrect use of agrochemicals. Thailand is reported to have little training on pesticide use and consequently farmers give little attention to the proper use of pesticides. Indonesian farmers use manual spray equipment in which 58 percent of the equipment leaks.[11]

An Ill-Fated Legislative Effort

Rigid laws exist concerning pesticide use in the United States, yet there are virtually no regulations for the exportation of banned or unregistered pesticides. Pesticide manufacturers spend years and millions of dollars testing their products before approval and registration. When these pesticides are not approved, U.S. manufacturers often export them to Third World countries with more lenient restrictions on pesticide use. As a result, twenty-six pesticide ingredients banned from use in the United States are exported to developing countries, and six of them are used in Mexico.[12]

The problem is twofold: first, toxicity threatens U.S. consumers in the "circle-of-poison" effect, in which unregistered or banned pesticides are exported to a developing country and sprayed on crops whose produce is then exported back to the United States. The EPA ranks pesticide residues as one of the leading health problems in the United States. A study conducted by the National Academy of Science estimated that in the next seventy years, one million additional cases of cancer in the United States will be caused by pesticide residues.[13]

In the early 1980s, 15 percent of beans and 13 percent of peppers imported from Mexico exceeded FDA limitations for pesticide residues. Recent FDA tests on imported foods reveal that contamination by illegal pesticides account for only 5 percent of imports; however, contamination rates are higher for imported carrots, pineapples, rice, peas, and pears. Moreover, the FDA only tests 1 or 2 percent of imports while the rest wind up in U.S. grocery stores.[14]

In June 1990, the U.S. Senate Agricultural Committee voted to ban the export of unsafe pesticides. The panel adopted this legislation as part of the 1990 farm bill and hoped that the House of Representatives would address the issue. Strong objection to the bill came from the National Agricultural Chemicals Association, a trade group consisting of pesticide manufacturers, whose 1989 export sales totaled $2.2 billion. The bill was never enacted and, although the issue continues to be debated, it has been largely ignored.[15]

Critics argue that most pesticide exports are merely unregistered in the United States rather than banned. Many pesticides formulated in the United States are never tested for approval because they are of no use to U.S. agricultural needs. Instead they are exported directly to countries with suitable soils or who grow produce that can utilize the chemicals. Critics also argue that, with a ban, countries will seek out other nations who are willing to supply the banned pesticides.[16]

It should be stressed that the exportation of banned pesticides to the Third World is not an isolated issue confined to the United States, for it has been documented that many European countries also export banned pesticides.

Banned Pesticides—A Complex Picture

Most countries now operate a pesticide registration program, or a "positive" list of pesticides allowed to be used, though implementation frequently presents problems. A general perception prevails that many pesticides are banned in Europe, North America, or other industrialized countries, and then exported to developing countries. In fact, a relatively small number of pesticides are completely banned. An analysis of the regulatory actions taken in Costa Rica, Tanzania, and Vietnam and the European Union against the thirty-nine pesticides identified as most targeted indicates a more complicated situation. For example:

- Aldicarb is banned in Tanzania and not registered in Vietnam, but is still registered in ten EU countries and in Costa Rica.

- Monocrotophos, parathion, and parathion methyl are registered in many European countries, but all are banned, severely restricted, or not registered in Tanzania, Vietnam, and Costa Rica.

- Chlorobenzilate, chlorpropham, fluoroacetamide, and 2,4,5-T are still registered in some EU countries, but are banned or not registered in Tanzania, Vietnam, and Costa Rica.

- Aldrin, chlordane, DDT, EDB, dieldrin, heptachlor, and HCH are banned in the EU, Costa Rica, and Vietnam, but Tanzania allows restricted or severely restricted use of these products.[17]

A global ban on some pesticides is urgently required, but will inevitably move slowly and cover a limited number (at present only the nine POPs pesticides may find agreement for a global ban and phase out on production and use) and even among these some exemptions may apply. Developing nations need good information about regulatory actions taken by governments with more resources to assess pesticides. It is important that they receive help in developing the capacity to implement regulation—including the ability to prevent import of pesticides that they have banned—and access to more and safer alternatives.

The Picture in Asia

More than $30 billion is spent on pesticides annually. A quarter of this total is spent in Asia, where sales increased by more than 10 percent in 2000. Thailand is the biggest spender in the South Asia Region, with pesticide sales equaling $247 million. Across Asia, however, there are more than 800 million people living in poverty. Out of desperation, farmers trust the sellers and promoters of the chemicals, those convincing them that pesticides will keep insects and weeds from destroying their crops, often the farmers' only means of income.

The Asia Crop Protection Association (APCPA), which represents such multinationals as Bayer, Cyanamid, Dow AgroSciences, DuPont, Novartis, and Zeneca, claim their products are reducing famine by minimizing crop damage by insects and weeds, and that they are saving lives through controlling disease-carrying insects. The global pesticide market is dominated by ten companies, which between them take 80 percent of more than $30 billion worth of sales.

There are many loopholes in the regulatory system. According to European legislation, only end products permitted in Europe can be exported. However, it is legal to export the starting product, the active ingredient of which is then manufactured into the end product in developing countries.

In places like Cambodia, struggling to rebuild its society after decades of civil war, the government is unable to regulate the flow of pesticides. Corporations such as the German company Bayer say it is their policy not to export dangerous chemicals to countries lacking proper regulation. Bayer also claims it abides by the laws of the importing country and ensures that it does not export products that are outlawed in those countries. However, one may ask, how are banned category 1a chemicals still available across Asia? Evidently conditions already described in most developing countries make it practically impossible to guarantee appropriate pesticide usage.

Seventy-three percent of imports into Thailand are WHO categories 1a and 1b, extremely toxic and highly toxic. In Cambodia, 84 percent of pesticides are moderately to extremely hazardous to human health. In developed countries these chemicals are either banned or they can be used only by licensed specialists, who must enforce a number of stringent precautions. In Southeast Asia, however, these chemicals are freely used without precautions. Labels, along with being written in a foreign language, fail to provide data on the active ingredient, application, date of manufacture, or safe handling of the chemical.

Methyl parathion is officially banned or restricted in Cambodia, China, Japan, Malaysia, Bangladesh, Indonesia, and Sri Lanka. However, in some Asian countries, it is widely used on a frequent basis. Folidol, Bayer's brand name for methyl parathion, is possibly the most popular insecticide on the Cambodian market. Cambodia has more than fifty kinds of dangerous pesticides: organophosphorus compounds such as methyl parathion, mevinphos, methamidophos, and moncrotophos are being illegally exported to Cambodia through Thailand and Vietnam. Cambodia serves as a dumping ground for products that cannot be sold in its neighboring countries. The multinational firms that manufacture the chemicals claim that they are not responsible because they do not directly market to Cambodia. Methamidophos, a WHO category 1a, can be fatal if swallowed, inhaled, or absorbed through the skin. Manufactured by Bayer and marketed as Monitor, methamidophos is a restricted chemical in the United States and New Zealand, but is still a favorite of Cambodian farmers.

Product Stewardship

Major manufacturers say they try very hard to encourage responsible use of the chemicals; they call it "product stewardship." There have been a number of global industry initiatives—including the Safe Use Campaign and the Responsible Care Initiative. Both aim to raise standards of understanding and practice throughout the distribution chain, from production to disposal. However, the highly toxic nature of some of the chemicals and conditions for users in developing countries render both of these initiatives inadequate. If international efforts to control pesticides are to have a significant impact then governments will need to start agreeing on targets and strategies to reduce pesticide use and to invest in sustainable pest control methods like IPM.[18]

The pesticide industry should be held responsible not only for their exports, but also for the way their products are used. Chemical companies say that it is not their responsibility if there are lax safety conditions in the countries that use their products. The multinationals blame small regional producers making generic versions of their products with little to no safety training and also resellers who smuggle products over the border from Thailand. While officials and corporations argue about who is responsible, pesticides continue to flow, poisoning millions of farmers, their families, and their environment.

The Costa Rican Experience

One of the most striking aspects of Central America's pesticide tragedy is the improvements that could be attained if the responsible parties were willing to take relatively modest corrective steps.

Costa Rica's regulatory system is widely viewed as the most advanced in the region, but this is not much of a claim. The country has a tiny, inadequately trained, and poorly equipped staff to oversee pesticide use. To compensate for its lack of toxicologists and biologists, Costa Rica, like its neighbors, looks to the north for guidance. Because many U.S. and European companies have exported pesticides that they are not allowed to sell at home, Costa Rican law required imported pesticides be approved for use in the country of origin. In practice, however, what sometimes passes for the country of origin is nothing more than a trans-shipment point. A pesticide banned for use in the United States can be shipped to an intermediary country to circumvent these restrictions. For example, U.S.-made haloxyfop has entered Costa Rica through Colombia.

Because they lack resources, Central American regulators typically resort to copycat pesticide restrictions. If the EPA bans a pesticide, for instance, it catches the attention of Costa Rican regulators, especially if it is a pesticide that the U.S. FDA spot-checks at the border. For example, U.S. regulators rejected shipping containers of Costa Rican produce for having illegally high residues of aldicarb. But these regulators do not check for many dangerous U.S.-made pesticides used in Costa Rica.

The 1991 incident just described shows that Costa Rican regulators can crack down on a pesticide almost overnight. There is something inexplicable, however, in the fact that they will do so in response to market pressure but not in response to the death of Costa Rican workers.

Another serious problem involves pesticides not prohibited outright in the United States, but subject to strict controls. These controls typically get lost in translation to the developing world. This is the case with many dangerous pesticides that, typically under industry pressure, are authorized for restricted use in the United States by specially trained workers who must follow specified safety conditions. While these stipulations and conditions may provide some measure of protection within the United States, they keep the door open to abuse in the developing world, where the stipulated safety requirements are not met in practice. Monsanto's alachlor (Lasso), for example, poses such an elevated cancer risk that it can only be applied in the United States by workers operating from inside sealed cabins. Such equipment is nonexistent in all but a few Costa Rican plantations.[19]

The Human Costs

Up to date global estimates are lacking, but there are 1.2 billion agricultural workers worldwide and it is likely that millions of pesticide poisoning cases still occur each year. In 2000, Brazil's Ministry of Health estimated the country had 300,000 poisonings a year and 5,000 deaths from agricultural pesticides, many of them imported.[20] In an Indonesian study, 21 percent of spray operations resulted in three or more neurobehavioral, respiratory, and intestinal signs of symptoms.[21] In a United Nations survey, 88 percent of pesticide-using Cambodian farmers had experienced symptoms of poisoning.[22]

Many of these problems are precisely those that the FAO Code and the PIC provisions were designed to minimize. Yet the pesticides implicated in these poisonings often remain outside the international regulatory systems.

Specific Poisoning Cases

In October 1991, it was reported that 350 people, thirty-one of whom died, were poisoned by endosulfan in Sudan when they ate bread made from contaminated maize flour. The manufacturers, Hoechst, argued that although endosulfan has high acute toxicity, its hazards can be overcome if the compound is used with proper care. Yet in 2000, a report from Benin showed the havoc that endosulfan is still causing to farming communities.

In June 1992, research compiled in Central America and Malaysia demonstrated how paraquat under conditions of use in these countries is a major occupational hazard for plantation workers. In June 1996, a study estimated that in the previous year there were 15,300 cases of pesticide poisoning in China, 91 percent involving organophosphate insecticides. In March 2000 a report on the Del Monte Kenya pineapple plantation revealed that pesticides recommended by the FAO as too dangerous to use in developing countries are used on the plantation. There appeared to be no training for workers, yet the company claims to take part in a program for the responsible use of pesticides run by the Global Crop Protection Federation.[23] Another account describes how Africans fishing on Ghana's Lake Volta discovered that if they dumped the insecticide Gammalin 20, imported for use by cocoa farmers, into the lake, many fish died and floated to the top for easy retrieval. These fish were then eaten by villagers or sold. The people began suffering dizziness, headaches, vomiting, and diarrhea—the first symptoms of poisoning by lindane, the active ingredient in Gammalin 20. Convulsions, brain disturbances, and liver damage followed. The fish population declined by up to 20 percent. The fishermen did not link the pesticide to the damage done to their health and their fishing until a private aid agency noted the connection.[24]

Corporate Accountability

During the 2003 annual meeting season, shareholders demanded that two of the three largest agrochemical companies acknowledge the environmental and health risks of their products. Led by socially responsible investment firms, shareholders at Bayer and Monsanto requested detailed information about the handling of dangerous pesticides and by-products and the possibility of costly lawsuits. Although not legally required, this information helps investors to assess their risk and promotes a corporate commitment to environmental health and safety. Bayer shareholders, meanwhile, highlighted the board's insufficient response to a tragic poisoning in Peru.

Investors at Monsanto's annual meeting in April 2003 expressed concern about the company's handling of unregistered carcinogenic and obsolete pesticide stocks. A resolution submitted by Harrington Investments, Inc., called on Monsanto to disclose its policies and procedures for exporting probable or likely carcinogens and pesticides not registered in the United States to developing countries. Since training and safety equipment are often limited or unavailable in these countries, Harrington Investments also requested disclosure of training and educational information Monsanto provides to farmers and farmworkers using these dangerous pesticides. The resolution garnered support from 13.32 percent of the voting shareholders.

German-based Bayer was the other company to come under fire as Luis Gomero of the Pesticide Action Newark–Peru appeared before the board in late May 2003 to demand justice for the victims of the 1999 Folidol (methyl parathion) poisoning in Tauccamarca, Peru. Folidol, a pesticide produced by Bayer, killed twenty-four schoolchildren and badly poisoned eighteen others after it was mistaken for milk powder at a local school. Bayer had marketed the pesticide, a white powder with no strong odor, in small plastic bags labeled in Spanish and without any appropriate pictograms to indicate its use or danger. The Spanish text was of little help to local farmers, most of whom speak Quechua dialect and are illiterate. Citing these failures, a Peruvian congressional subcommittee found Bayer criminally responsible for the poisonings in 1999.

In order to ensure justice for the Tauccamarca victims, Gomero would like the Dow board to accept responsibility for the poisoning; provide medical monitoring, care, and special education for the surviving children as necessary; establish a functioning health post in the village; and recognize the families' suffering, in part through financial compensation.

In his response to Gomero's testimony, Bayer chairman Werner Wenning asserted that the pesticide that poisoned the Tauccamarca children was not a Bayer product, and the Peruvian court dismissed any claims against Bayer. However, representatives of the Tauccamarca families pointed out that Bayer had registered both pesticides implicated in the case (methyl parathion and ethyl parathion) for use in Peru, and the Peruvian court had not received all relevant documents and had not ruled in the case.

Gomero's appearance was coordinated by the German group Coordination gegen BAYER-Gefahren (CBG, or the Coalition Against Bayer Dangers) using Bayer shares held by Pesticide Action Network Germany. CBG has been working since 1978 to increase Bayer's transparency, publicize its global abuses and violations, and ensure appropriate responses and compensation.

The impact of these recent shareholder actions remains uncertain, but as John Harrington reminds investors: "Companies with greater corporate responsibility and transparency prosper long term."[25]

Arguments Pro and Con

Pesticide manufacturers in the United States generally argue that additional restrictions on exports are unnecessary because they claim they do not manufacture and export pesticides that have been denied registration in the United States. They contend that more stringent export controls for unregistered pesticides (for example, pesticides for which U.S. registration was not sought) could unfairly prohibit the export of products for which there is little evidence of environmental risk. Pesticides may not be registered simply because manufacturers do not wish to sell them in the United States, and some may be approved by regulatory agencies in other countries. Other pesticides may not be registered for economic or marketing reasons or because target pests are not a problem on crops grown in this country. The global nature of pesticide production and distribution further complicates the issue: a U.S. law cannot prevent the manufacture and use of pesticides in other countries. U.S. manufacturers might simply relocate production facilities, or production by foreign manufacturers might increase.

Environmentalists argue that any pesticide product not registered in the United States has not been approved by the EPA and is potentially unreasonably harmful. Thus, they support proposals to prohibit exports of unregistered pesticides as a means of protecting the global environment. They also believe it would protect American consumers from unsafe pesticide residues on imported foods.[26]

Concluding Observations

In the Third World, often both adults and children are involved in the application of pesticides; many mix pesticide formulations with their hands and must work the fields in bare feet. In addition to exposure during the direct application of pesticides, farmers and agricultural workers face exposure when they re-enter sprayed fields for crop management and harvesting activities. Moreover, contamination of water sources, proximity to aerially sprayed fields, inadequate storage facilities, and the reuse of pesticide containers can affect entire families or communities. Not even unborn children are safe: exposure to chemicals, especially endocrine disrupters, during fetal development can cause permanent damage.

For farmers in the tropics, fully protective garb is too hot and costly to maintain; farmers there accept illness as a necessity. Integrated Pest Management has previously been demonstrated to reduce pesticide use with no loss of crop yield. The frequency of spraying should be reduced through widespread training in IPM.[27]

Governments in developing countries need to invest more in the skills required to interpret scientific and technical data and use it to make sound local risk assessments and to implement regulations. Resources for raising awareness are equally crucial: most users of pesticides in developing countries not only have limited perceptions of the risks, but also a high acceptance of risks due to competing priorities essential for survival.

The most effective controls over pesticides are good national registration schemes that require tests backed by solid information appropriate for the local conditions. Each different pesticide mixture or formulation should be registered for use on each crop on which it is intended to be used. Aided by the FAO, many developing countries have now introduced pesticide registration programs. The problem is they do not have the capacity to implement these regulations.

Many developed countries permit the export of pesticides that are banned, restricted, or unregistered within their own borders. This practice raises many ethical issues as well as economic, social, political, and public health issues.[28]

Corporations and their executives, accustomed to getting away with exporting domestically banned pesticides and causing detrimental health and environmental effects in developing countries, will change only if the costs of this activity outweigh the benefits. Multinational corporations have the power to co-opt Third World governments, and only a strong international code which is well enforced and costly to disobey can prevent these types of unethical practices. Those injured by hazardous pesticides will not be protected from an alliance of corporations seeking to continue exporting domestically banned pesticides in the name of free trade unless citizens organize and demand the WTO not allow corporations to export domestically banned pesticides as prohibited by the PIC agreement of the FAO. The fight for corporate responsibility will not be a top-down fight but a bottom-up struggle.

Notes

1. Professor Hermann Waibel, briefing for the Developing Countries Project, Pesticides Policy Project, adviser to the Global IPM Facility.

2. Michael Holly, "The EPA's Pesticide Export Policy: Why the United States Should Restrict the Export of Unregistered Pesticides to Developing Countries," *New York University Environmental Law Journal* 9 (2) (2001): 340–385.

3. Al Krebs, "U.S. Exports More and More Poison Pesticides," *Agribusiness Examiner* 142 (February 4, 2002).

4. *Early Warning System on Global Trade in Hazardous Pesticides and Chemicals to Become Law* (Penang, Malaysia: PAN-Asia-Pacific press release, n.d.)

5. Ibid.

6. Janet Raloff, "The Pesticide Shuffle," *Science News* 149 (11) (March 16, 1996): 174.

7. Ibid.

8. *Prevention and Disposal of Obsolete Pesticides* (Geneva: Foreign Aid Organization Pesticide Disposal Series, n.d.)

9. *Pesticides in Your Food* (London: Pesticides Action Network-U.K., n.d.)

10. Douglas L. Murray and Peter Leigh Taylor, "Claim No Easy Victories: Evaluating the Pesticide Industry's Global Safe Use Campaign," *World Development* 28 (10): 1,735–1,749.

11. Mike Owen, "Unsafe Pesticide Use," *Weed Science* (Ames, IA: Iowa State University, July 1, 1997).

12. David J. Hanson, "Administration Seeks Tighter Curbs on Exports of Unregistered Pesticides," *Chemical and Engineering News* 72 (7): 16–17.

13. Richard Tansey, "Eradicating the Pesticide Problem in Latin America," *Business and Society Review* 2 (Winter 1995): 55–59.

14. Cameron Barr, "Combating the 'Circle of Poison,'" *Christian Science Monitor*, May 7, 1991.

15. Bruce Ingersol, "Senate Panel Targets 'Circle of Poison' as It Votes to Ban Some Pesticide Exports," *The Wall Street Journal*, June 7, 1990.

16. Frank Edward Allen, "Lines Are Drawn Again Over Pesticide Exports," *The Wall Street Journal*, May 28, 1991.

17. *Pesticide Hazards* (Seattle, WA: Ecochem, Inc., February 6, 2005).

18. D. Lorring, "Struggling to Keep Cambodia off the Pesticide Treadmill," *Global Pesticide Campaigner* 5 (4) (December 1995).

19. Andrew Wheat, "Toxic Bananas," *Multinational Monitor* 17 (8) (September 1996).

20. N. Bensugan, "Agritoxicas: situacao extramanenta grave pode plorar ainda mais," *Noticias Socioambientias* (Rio de Janeiro), 2000.

21. M. Kishi et al., "Relationship of Pesticide Spraying to Signs and Symptoms in Indonesian Farmers," *Scandinavian Journal of Work, Environment, and Health* 21 (1995): 124–133.

22. P. Sodavy et al., *Farmers' Awareness and Perception of the Effects of Pesticides on Their Health* (Geneva: FAO Community IPM Programme Field Document, April 2000.

23. Topsy Jewell, "International Issues Covered by *Pesticide News* in its Fiftieth Issue," *Pesticide News* (December 2000).

24. Global Tomorrow Coalition, *The Global Ecology Handbook: What You Can Do About the Environmental Crisis* (Boston: Beacon Press, April 1990).

25. Coalition Against BAYER Dangers, "Agrochemical Shareholders Call for Corporate Accontability," *Global Pesticide Campaigner* (August 2002).

26. Linda-Jo Schierow, *95016: Pesticide Policy Issues* (Washington, D.C.: CRS Report for Congress, December 4, 1996).

27. Kishi et al. Op. cit.

28. L. K. Lowry and A. L. Frank, "Exporting DBCP and Other Banned Pesticides: Consideration of Ethical Issues," *International Journal of Occupational and Environmental Health* 5 (2) (April–June 1999): 135–141.

Eight

Remedies and Reflections

We should no longer accept the counsel of those who tell us that we must fill our world with poisonous chemicals; we should look about and see what other course is open to us.

—Rachel Carson, *Silent Spring*[1]

Necessary Policy Reforms

The newly discovered connections between pesticides and disease just begin to scratch the surface of the potential impact of chemicals on public health. Tens of thousands of industrial chemicals on the market have not been tested for developmental health effects at low doses. No public health information exists for close to half of the high production-volume pesticides. Moreover, where significant evidence of harm to public health already exists, inadequate resources and legal authority often prevent regulatory agencies from taking preventative actions.

In order to protect the public from toxic exposures, we must take firm steps to remedy the ignorance about health effects of widely used pesticides and empower regulatory agencies to ensure that consumer products do not contain dangerous pesticides. These steps include:

1. Phasing out pesticides that persist in the environment, accumulate in organisms, or for which evidence of potential harm to human health exists from exposure.

2. Requiring pesticide manufacturers to develop analytical techniques to detect the chemicals they produce, and relevant breakdown products, in the environment and organisms, and to submit these techniques to the state. Taxpayers currently pay scientists to guess at what emerging pesticide threats may be present in our environment and bodies and then develop the testing methods to detect them. This causes significant delays in determining which pesticides pose the greatest threat to public health.

3. Requiring pesticide manufacturers to supply the state and federal government with toxicity data for their products, including low-dose effects (it has recently been revealed that low-dose exposures of certain pesticides are more dangerous than those at higher doses) on development and reproduction.[2]

Policy makers are faced with what to do about suspected toxins when there is uncertainty or ambiguity in the science used to judge risk. Industry members continue to argue that it is irresponsible to sacrifice new products and undermine fiscal prosperity by halting product development before the data conclusively indicate danger. The precautionary principle should prevail: when society is faced with devastating health problems as a result of using potentially toxic chemicals, those chemicals should be held in abeyance until they are proven safe.

The Precautionary Principle

One of the most quoted definitions of the precautionary principle is the Wingspread Statement, produced by a gathering of scientists, philosophers, lawyers, and environmental activists in the United States in 1998. It pronounces that "when an activity raises threats of harm to the environment or human health, precautionary measures should be taken even if some cause-and-effect relationships are not fully established scientifically."[3]

The substance of the precautionary principle is not really new. The essence of the principle is captured in such cautionary clichés as "An ounce of prevention is worth a pound of cure," "Better safe than sorry," and "Look before you leap." The precautionary principle may be interpreted as a generalization of the ancient medical principle associated with Hippocrates: "First, do no harm." The essence of the principle is the idea that if the consequences of an action are potentially severe or irreversible, the absence of full scientific certainty should not be used to prevent action. In other words, the onus should be on pesticide manufacturers to demonstrate beyond reasonable doubt that their products will not harm people or wildlife before these products are approved for use.

Cosmetic Pesticides Are Harmful

In the United States, we are enthralled by the sight of a picture-perfect lawn or garden, for which credit is given to the application of large doses of dangerous pesticides. Canada is head and shoulders above us in their treatment of these chemical threats. More than seventy Canadian municipalities, including Toronto, Quebec, and Halifax, have taken action to reduce or ban cosmetic (or "aesthetic") use of pesticides on both public and private property. Most of these bans are being phased in over a few years. The sales of pesticides have not been completely banned, though. For this reason, other cities, such as Vancouver, have opted against a legislative approach and instead are trying to educate consumers on why they should not use pesticides and to try safer alternatives instead. Canadian public health officials are working to keep pesticides off lawns and gardens.[4] We could do worse than to emulate their actions.

Multiple Chemical Sensitivity

Another area that does not receive the attention it deserves concerns the plight of persons who are afflicted with multiple chemical sensitivities. These people include those with asthma or allergies, as well as individuals with chemical sensitivities who suffer the effects of pesticide exposure more severely than those without. But how do some people become so sensitive to begin with? Is it because of multiple exposures they've received throughout their fetal development and adult lives? For such people, day-to-day living can present challenges, for there is often nowhere to hide from the widespread and persistent use of these toxins.

Changes in Agriculture

In Chapter Two, pesticide use in U.S. agriculture was examined since it is the locus of the heaviest pesticide utilization. Many changes have occurred in agriculture in the past few decades. During the 1950s and 1960s American farmers depended on cheap energy, plentiful water supplies, and extensive use of chemical fertilizers and pesticides to produce high yields with decreasing labor on reduced amounts of land. In recent years the costs of fuel and chemicals have increased sharply, the high use of pesticides has led to resistance in many pest species, and concern has developed over environmental contamination from fertilizers and pesticides. Increasing attention, therefore, is being given to means of reducing the reliance of American farmers on highly chemical means of production. To produce high yields, protect soil productivity, and maintain environmental quality, farming must be based on an understanding of how water and dissolved chemicals move through the plant-soil-groundwater system. A small but growing percentage of farmers are farming with no pesticides, and many others are reducing their overall chemical use. Agricultural research has begun to focus on ways of maintaining environmental quality while producing acceptable crop yields. One example is Integrated Pest Management, aimed at controlling pests through a combination of methods that minimize undesirable ecological effects. Continuing research and education need to be conducted on farming practices that produce profitable yields while maintaining environmental quality and the long-term productivity of the land.

A Long Way To Go

Opposition to the use of pesticides has been most successful in wealthy countries such as Canada and the United States. Many communities have passed restrictive legislation, mostly pertaining to cosmetic use of pesticides. While this has been rightly hailed as a major step forward for the environmentally sensitive and for the protection of community health, there remains the larger question of how to protect the young and the poor from excessive exposures. While it's nice that alternatives are being promoted, there is still a long way to go until the public knows the hazards of pesticides. While tanker trucks spewing clouds of poisons are an obvious menace, the larger issue of home and institutional treatments has not been so widely discussed.

The opposition to indiscriminate use of pesticides brings up many questions about our role in nature, and what exactly constitutes a pest. As a new generation confronts the greater problems of ecological degradation, pesticide use will continue to attract activists. The rest of us owe these people a debt of gratitude, because like all responsible, ecologically minded individuals, they are taking charge of the health and well-being of our communities. They refuse to accept the assurances of public officials that public health is being protected. They also refuse to accept the belief that we can do whatever we want with nature with no regard for the consequences.

IPM and Sustainable Agriculture

It is important to note that farmers generally use pesticides very judiciously. Chemicals are one of the most expensive inputs that a farmer can use. IPM and sustainable agriculture provide alternative technologies that allow farmers to reduce pesticide usage while maintaining productivity and profitability. IPM integrates all pest management techniques into one crop management strategy. Pesticides may be used to control a pest only when the pest is threatening economic losses to a crop. IPM programs rely on biological control, scouting of crops, and other cultural practices as well as reduced chemical inputs.

Sustainable agriculture is farming practices that preserve and protect the future productivity and health of the environment. Sustainable agriculture is, however, a broader topic than organic farming. The way food is processed, packaged, and transported may pose a threat to the environment, even when the food was cultivated organically. For example, pretzels may be organic—meaning 95 percent of their ingredients are organically grown—but in reality, they have been produced from highly refined flour processed using energy-wasting machinery, packaged in non-recyclable plastic, and shipped around the world using large amounts of fossil fuel. Growing foods organically is, therefore, only the first step in achieving sustainable agriculture. Most environmentalists and ecologists and many individuals involved in the production of organic foods believe that sustainable agriculture is necessary if we are to reach the long-term goals of personal health and ecological balance.

Organic Agriculture

In the United States, organic agriculture is expanding, while conventional agriculture is in decline or at best stagnant. But the organic method of farming is still viewed with suspicion by many U.S. farmers who are used to getting advice from chemical company salespeople or from university agriculture departments that are heavily endowed and influenced by chemical companies and bioengineering firms.

Unfortunately most organic certifiers and other pro-organic organizations don't have well-developed programs for encouraging conventional farmers to make the transition from chemical to organic farming. As a result, many farmers don't have the information they need to evaluate whether going organic makes sense for them. The USDA's Coop Extension Service should be encouraging the transition to organics, but it is not. In the entire USDA bureaucracy there is only one extension agent who is a qualified expert in organic agriculture. Predictably, he's located in Santa Cruz, California.

Farming is hard work, and in many cases farming organically is even harder. Weeds, harmful insects, fungi, and other pests must be monitored and managed much more painstakingly, while beneficial insects and soil organisms are encouraged. Organic growers must keep up with research about alternative methods of pest control and apply the results on their farms. They must learn to control pests using an ecosystem approach that requires patience because sometimes it takes several years to produce satisfactory results. Even more important, organic farmers must care for and enrich their soil with organic matter and non-chemical sources of the nutrients their crops require. This is often much more labor intensive that the yearly applications of fertilizers and pesticides that conventional farmers use.

But organic agriculture is viable and practical for most crops, and farmers who are motivated can, without too much trouble, find the information and resources they need to successfully transition more farm operations—such as grains, vegetables, livestock, eggs, and herbs—to organic methods. In fact, as the success of the organic agriculture movement has proven, synthetic chemical pesticides are probably far less necessary than we have been led to believe by their producers. Moreover, the reality is such that chemical pesticide use is almost always harmful to someone or something, whether farmworkers, consumers, non-target organisms, or the rest of the environment.[5]

The Body Burden

Toxic chemicals, both naturally occurring and man-made, often enter the human body. We may inhale them, swallow them in contaminated food or water, or, in some cases, absorb them through our skin. A woman who is pregnant may pass chemicals to her developing fetus through the placenta. The term "body burden" refers to the total amount of these chemicals that are present in the human body at a given point in time. Sometimes it is also useful to consider the body burden of a specific, single chemical, such as, for example, lead, mercury, or dioxin.

Some chemicals or their breakdown products (metabolites) lodge in our bodies for only a short time before being excreted, but continuous exposure to such chemicals can create a persistent body burden. Arsenic, for example, is mostly excreted within seventy-two hours of exposure. Other chemicals, however, are not so readily excreted and can remain for years in our blood, adipose (fat) tissue, semen, muscle, bone, brain tissue, or other organs. Chlorinated pesticides such as DDT can remain in the body for fifty years. Whether chemicals are quickly passing through or are stored in our bodies, body burden testing can reveal an individual's unique chemical load and can highlight the kinds of chemicals we are exposed to as we live out each day of our lives. Of the tens of thousands of chemicals that are used in the United States, we do not know how many can become a part of our body burden, but we do know that several hundred of these chemicals have been measured in people's bodies around the world.

Scientists estimate that everyone alive today carries within her or his body at least 700 contaminants, many being pesticides, most of which have not been well studied.[6] This is true whether we live in a rural or isolated area, in the middle of a large city, or near an industrialized area. Because many chemicals have the ability to attach to dust particles and/or catch air and water currents and travel far from where they are produced or used, the globe is bathed in a chemical soup. Our bodies have no alternative but to absorb these chemicals and sometimes store them for long periods of time. Wherever we live, we all live in chemically contaminated neighborhoods.

The fact that we have residues of hundreds of industrial and agricultural chemicals in our bodies is a direct invasion of our most private property. Pesticides that present serious health risks should simply be banned from use. Under such a policy, research would focus only on detecting those pesticides that cause such problems. This would eliminate the task of developing tolerance levels and eliminate risk rather than attempting but failing to manage it. Effective alternatives currently exist for most pesticides. The social costs of continuing their use outweigh the short-term economic gains they provide to the pesticide and food companies.

Although organic food is more expensive than crops grown with pesticides, the external costs of pesticides are not included in the price of commercially grown food. It has been very roughly estimated that a direct investment of $4 billion in pesticides saves about $16 billion in crop losses, but causes an estimated $8 billion in environmental and health costs to society. Five billion dollars of that is paid for by society and not by pesticide manufacturers or direct users.[7] Buying organic food from local farms not only enhances our own health and that of future generations, but also benefits us as it decreases pollution; supports local, small-scale farmers; and makes farming itself more sustainable in the long run.

Pesticides have been in existence for only about sixty years. Gardeners, farmers, and foresters have always had to control pests using methods such as crop rotation, companion planting, and biological controls. Pest management has only recently become virtually synonymous with the use of pesticides. We must reverse this unsustainable trend.

Pesticide manufacturers should be required to submit safety data that cover all the likely combinations affecting the human body. Safety data should be required on possible impacts on the most vulnerable—the elderly, infants, and young children. Safety data should specifically cover individuals most likely to receive exceptionally high does of pesticides and similar chemicals, such as bystanders who are also farmworkers, those who also use pesticides in their gardens, and those who do not peel their fruit.

Protect Our Children

Children must be better protected from both new and existing pesticides that are known or possible toxins. To protect children from existing toxins, the EPA and FDA need more authority and resources to regulate and reduce emissions and exposures. Under the current system, efforts to tighten regulations to protect children from known toxins are costly and protracted. Indeed, countless communities across the globe suffer from widespread environmental contamination. If there is any lesson from our experience with environmental toxins, it is that we need to identify pesticides that are toxic before they are marketed or widely disseminated.

For all new chemicals, including pesticides, extensive pre-market testing should be required in multiple animal species of both sexes and at different developmental stages. These tests should be designed to have adequate statistical power to detect subtle differences within the ranges of exposure that occur in human populations. If implemented, these testing requirements would represent a dramatic departure from existing regulations, while providing a powerful incentive for industry to develop less-toxic chemicals.

Toxicity testing in animals is essential but insufficient to protect pregnant women and children. For one thing, uncertainties about the safety of a chemical for humans will persist even after toxicity testing in animals is successfully completed. One additional safeguard that deserves further debate is whether prevalent environmental chemicals to which children could be exposed should undergo more extensive testing in human trials before they are marketed. If done, these trials should examine exposure, uptake, and adverse effects among children or other populations only when the product is used as intended. For example, once animal toxicity testing of a residential pesticide is complete, including developmental neurotoxicity and reproductive toxicity testing, a pesticide could undergo further testing in the home environment. Using an experimental group and a control group, researchers would compare levels of pesticides found in settled dust, on children's hands, and in their blood, urine, or hair. Children would be followed, when indicated, to ensure that an excess of neurobehavioral problems or other relevant outcomes did not develop among those whose homes received pesticide applications.

The Pesticide Applicator's Plight

One important fact should not be overlooked. Pesticides are widely used on urban landscapes as well as in agriculture. Consequently, pesticide applicators are on the front lines, facing multiple potential routes of exposure to these toxic chemicals. While the use of protective equipment is important, even gloves, masks, and full-body protective clothing do not completely eliminate exposure.

Pesticide exposure can cause lost workdays for an applicator, or, in the worst cases, permanent injury, disease, or even death. Exposure can also affect the health of an applicator's offspring, and even low-level exposures can cause harm. Other family members can be exposed to residues brought home on work shoes and clothing.

For the sake of their own health and that of their children and families, applicators should want to find ways to reduce their exposure to pesticides. The most effective way to reduce exposure is to avoid the use of pesticides. Fortunately, safer methods are available for controlling many pests. It is important for applicators to educate their employers or clients about the hazards of pesticides and encourage them to use non-chemical pest control methods.

It took decades for science to establish a firm causal link between smoking and lung cancer. It will undoubtedly take many more years for research to conclusively demonstrate links between pesticide exposure and cancer and other chronic diseases. But enough is known now to raise strong suspicions, and the acute exposure hazards are clear. Why take chances with people's health?

Public Consciousness Is Awakening

Pesticides are used worldwide in agriculture, industry, public health, and domestic applications; as a consequence, a great part of the population may be exposed to these compounds. In spite of this extensive use, knowledge of the health risks associated with prolonged exposure is rather poor, and major uncertainties still exist. Epidemiological observations in man have so far produced little conclusive information, mainly because of weaknesses in exposure assessment.

Pesticides are the only class of toxic materials intentionally introduced into the environment to kill or damage living organisms. Currently, people are exposed without their knowledge to pesticides whose human health effects are largely unknown. To protect our health and help safeguard our water and food from contamination, we need better information about pesticide use, whether on golf courses, at schools, in homes, on farms, or on suburban lawns and gardens.

Being educated about pesticides is a basic right. Knowing what toxic substances are in one's environment is a matter of fundamental fairness and is an essential part of a democratic society. Information about pesticide use can help individuals make choices and take action to limit their exposure.

It is extremely difficult, if not impossible, to prove that a particular chemical caused a person to become chemically sensitized, a tumor to form, a miscarriage to occur, or brain damage to happen. After all, we live in a society that does not tolerate dangerous experimentation on humans. However, based on the growing body of evidence from laboratory research, wildlife studies, and accidental human exposures, it is very clear that currently used chemical pesticides pose serious threats to human health.

Pesticide reporting data once again confirm our addiction to these hazardous chemicals. These findings arrive at a key point in time, when a steady drumbeat in the media has raised public consciousness of pesticide risks, prompting new questions from an ever-broader cross-section of citizens and policy makers. As never before, people are recognizing that pesticides are not silver bullets, but clumsy, non-specific poisons that leave an inevitable trail of contamination in their wake and do predictable harm. Continued reliance on pesticides puts citizens at unnecessary risk. The time is ripe for our policy makers to reverse course, to reject the risks and financial burdens foisted upon society by pesticide manufacturers—who employ battalions of lobbyists and a vast public relations machine to impede reform at every level—and make pesticide alternatives the norm.

The tide is turning. The public is becoming better informed and taking control of its health and food. New pesticide-free farming and gardening methods are being announced. These new methods are what your grandparents and great-grandparents used. While the pesticide industry might have powerful lobbying firms in Washington, D.C., public education about this subject will ultimately prevail. History has taken us far past the point where we could envision living without some of the services synthetic pesticides provide, but it has also taken us to the point where we can no longer shrug off these warnings.

Notes

1. Rachel Carson, *Silent Spring* (Penguin Modern Classics, 2000).
2. "Growing Up Toxic: Chemical Exposures and Increases in Developmental Disease," *Environmental Health Reports* (June 23, 2004).
3. *Wingspread Statement on the Precautionary Principle*, Bette Hileman, *Chemical & Energy News*, (February 9, 1998): 16–18.
4. "Canada's Supreme Court Supports Pesticide Ban," *Environmental Science & Technology* (January 18, 2006).
5. "Organic Agriculture and America's Rural Crisis," *The Progressive Populist* (2000).
6. J. Onstot, R. Ayling, and J. Stanley, *Characterization of HRGC/MS Unidentified Peaks from the Analysis of Human Adipose Tissue. Volume 1: Technical Approach* (Washington: D.C.: U.S. Environmental Protection Agency, Office of Toxic Substances, 1987).
7. "Silent Spring II, Recent Discoveries Reveal New Threats of Pesticides to Our Health," *Third World Traveler* (Food First Newsletter: Summer 1997).

Selected Bibliography

Books

Briggs, S. A. *Basic Guide to Pesticides: Their Characteristics and Hazards*. Silver Spring, MD: Rachel Carson Council, 1992.

Carson, Rachel. *Silent Spring*. Boston: Houghton Mifflin, 1992.

Colburn, T., et al. *Our Stolen Future*. New York: Plume, 1997.

Matthews, Graham. *Pesticides: Health, Safety and the Environment*. London: Blackwell Publishing, Inc., 2006.

National Research Council. *Pesticides in the Diets of Infants and Children*. Washington, D.C.: National Academy Press, 1993.

Natural Resources Defense Council. *Intolerable Risk: Pesticides in Our Children's Food*. New York: Natural Resources Defense Council, 1989.

Needleman, H. L., and P. J. Landrigan. *Raising Children Toxic Free*. New York: Farrar, Strauss, and Giroux, 1994.

Repetto, R., and S. Baliga. *Pesticides and the Immune System: The Public Health Risks*. Washington, D.C.: World Resource Institute, 1996.

Schettler, T., et al. *Generations at Risk*. Cambridge, MA: MIT Press, 2000.

Solomon, Gina. *Trouble on the Farm: Growing Up with Pesticides in Agricultural Communities*. New York: Natural Resources Defense Council, 1998.

Steingraber, S. *Living Downstream*. New York: Vantage Press, 1998.

Wargo, John. *Our Children's Toxic Legacy: How Science and Law Fail to Protect Us from Pesticides*. New Haven, CT: Yale University Press, 1996.

Wiles, Richard, and Christopher Campbell. *Pesticides in Children's Food*. Washington, D.C.: Environmental Working Group, 1993.

Journal Articles

Alavanja, M. D., and S. Sandler et al. "The Agricultural Health Study." *Environmental Health Perspectives* 104 (1996): 362–369.

Chipman, H., and P. Kendall et al. "Consumer Reaction to a Risk/Benefit/Option Message about Agricultural Chemicals in the Food Supply." *The Journal of Consumer Affairs* 29 (1995): 144–163.

Davis, J. R., R. C. Brownson, and R. Garcia. "Family Pesticide Use in the Home, Garden, Orchard, and Yard." *Archives of Environmental Contamination and Toxicology* 22 (1992): 260–266.

Fenske, R., et al. "Potential Exposure and Health Risks of Infants Following Indoor Residential Pesticide Applications." *American Journal of Public Health* 80 (1990): 689–693.

Fernandez-Cornejo, J., and C. Greene et al. "Organic Vegetable Production in the U.S.: Certified Growers and Practices." *American Journal of Alternative Agriculture* 13 (1998): 69–78.

Fineberg, H. V. "Improving Public Understanding: Guidelines for Communicating Emerging Science on Nutrition, Food Safety, and Health." *Journal of the National Cancer Institute* 90 (1998): 194–199.

Fisher, B. E. "Organic: What's in the Name?" *Environmental Health Perspectives* 107 (1999): A150–A153.

Fleming, L. E., and J. A. Beal et al. "Mortality in a Cohort of Licensed Pesticide Applicators in Florida." *Journal of Occupational and Environmental Medicine* 56 (1) (1999): 14–21.

Flury, M. "Experimental Evidence of Transport of Pesticides through Field Soils—A Review." *Journal of Environmental Quality* 25 (1996): 25–45.

Gorrell, J. M., and C. C. Johnson et al. "The Risk of Parkinson's Disease with Exposure to Pesticides, Farming, Well Water, and Rural Living." *Neurology* 50 (1998): 1,346–1,350.

Gurunathan, S. M., and M. Robson et al. "Accumulation of Chlorpyrifos on Residential Surfaces and Toys Accessible to Children." *Environmental Health Perspectives* 106 (1998): 916.

Hill, R., S. Head, and S. Baker et al. "Pesticide Residues in Urine of Adults Living in the United States: Reference Range Concentrations." *Environmental Research* 71 (1995): 99–108.

Hoar, S. K., and A. Blair et al. "Agricultural Herbicide Use and Risk of Lymphoma and Soft-Tissue Sarcoma." *Journal of the American Medical Association* 256 (9) (1986): 1,141–1,147.

Hotchkiss, J. H. "Pesticide Residue Controls to Ensure Food Safety." *Critical Reviews in Food Science and Nutrition* 31 (1992): 191–203.

Lessenger, J. E., and M. D. Estock et al. "An Analysis of 190 Cases of Suspected Pesticide Illness." *Journal of the American Board of Family Practice* 8 (1995): 278–282.

Lowengart, R. A., and J. M. Peters et al. "Childhood Leukemia and Parents' Occupation and Home Exposures." *Journal of the American Cancer Institute* 79 (1987): 39–46.

Paul, M. "Occupational Reproductive Hazards." *Lancet* 349 (1997): 385–388.

Pennington, J. A. T. "The 1990 Revisions of the FDA Total Diet Study." *Journal of Nutrition Education* 24 (1992): 173–178.

Rea, William J. "Pesticides." *Journal of Nutritional and Environmental Medicine* 6 (1996): 55–124.

Roberts, J., and P. Dickey. "Exposure of Children to Pollutants in House Dust and Indoor Air." *Review of Environmental Contamination and Toxicology* 143 (1995): 59–78.

Sever, L. E., T. E. Arbuckle, and A. Sweeny. "Reproductive and Developmental Effects of Occupational Pesticide Exposure: The Epidemiologic Evidence." *Occupational Medicine: State of the Art Reviews* 12 (2) (1997): 305–325.

Whitmore, R. W., and F. W. Immerman et al. "Non-Occupational Exposures to Pesticides for Residents of Two U.S. Cities." *Archives of Environmental Contamination and Toxicology* 26 (1994): 47–59.

Zahm, S. H., and M. H. Ward. "Pesticides and Childhood Cancer." *Environmental Health Perspectives* 106 (1998): 893–908.

Zahm, S. H., M. H. Ward, and A. Blair. "Pesticides and Cancer." *Occupational Medicine: State of the Art Reviews* 12 (2) (1997): 269–289.

Government Reports

U.S. General Accounting Office. *Food Safety: Changes Needed to Minimize Unsafe Chemicals in Food.* Washington, D.C.: GAO, September 1994.

U.S. General Accounting Office. *Food Safety: USDA's Role Under the National Residue Program Should be Reevaluated.* Washington, D.C.: September 1994.

U.S. General Accounting Office. *National Survey of Pesticides in Drinking Water Wells; Phase I Report.* Washington, D.C.: EPA, 1990.

U.S. General Accounting Office. *Pesticides: Improvements Needed to Ensure the Safety of Farmworkers and Children.* Washington, D.C.: September 1994.

U.S. General Accounting Office. *Pesticides Industry Sales and Usage: 1996 and 1997 Market Estimates.* Washington, D.C.: Office of Prevention, Pesticides, and Toxic Substances, 733-R-99-001, 1999b.

U.S. General Accounting Office. *Pesticides: Use, Effects, and Alternatives to Pesticides in Schools.* Washington, D.C.: GAO/RCED-00-17, 1999.

Newspaper Articles

Eilperin, Juliet. "Study of Pesticides and Children Stirs Protests." *Washington Post,* October 30, 2005, A23.

Janofsky, Michael. "EPA to Bar Data from Pesticide Studies Involving Children and Pregnant Women." *New York Times,* September 7, 2005, A22.

Janofsky, Michael. "Limits Sought on Testing for Pesticides." *New York Times,* June 30, 2005, A23.

Kenworth, Tom. "EPA Caught Between Farmers, Food Safety Fears." *Washington Post,* August 2, 1999, A1.

Other Resources

Elderkin, S., et al. *Forbidden Fruit: Illegal Pesticides in the U.S. Food Supply.* Washington, D.C.: Environmental Working Group, February 1995.

Marquardt, C., C. Cox, and H. Knight. *Toxic Secrets: "Inert" Ingredients in Pesticides: 1987–1997.* Eugene, OR: Northwest Coalition for Alternatives to Pesticides, 1998.

Piper, Courtney, and Kagan Owens. *Are Schools Making the Grade? School Districts Nationwide Adopt Safer Pest Management Policies.* Washington, D.C.: Beyond Pesticides, 2002.

Index

Multiresidue methods (MRM): detection of different pesticide residues, 68; FDA testing of food shipments, 77

National Association of Farmworkers v. Marshall, 31

National Association of School Nurses: support of IPM, 114

National Association of State Departments of Agriculture (NASDA), 66

National Organic Standard, 103

National Parent-Teachers Association: support of IPM, 114

National Pesticide Telecommunications Network, 39–40

Nervous system: monitoring of significant impacts, 44

Non-detect factor, 85, 88

Notification laws: lawn pesticides, 176–77; posting, 177; registries, 177; state preemption of local laws, 177

Nutrient applications: unnecessary, 176

Occupational safety and health, 27–28, 28–29*f*; OSHA Field Sanitation Standard, 28–30; poor enforcement, 34

Off-target spray: air pollution and, 194

Ohio: State pesticide regulations for schools, 128

Organic agriculture, 238–39

Organic foods, 56–58; booming market for, 104; certification, 103; for children, 99–101; claimed benefits, 99; corporate inroads, 105–06; demand outstrips supply, 105; fewer pesticide residues, 102; fewer pesticide residues in foods, 78–79; increased antioxidant levels, 101–02; labeling, 103–04; reasons for growth, 98–99; standards, 106–07

Organic Foods Production Act of 1990, 107

Organic lawn care, 180

Organic matter: pesticide leaching and, 210

Organic standards: failure of a legislative challenge, 106–07

Organic Trade Association, 106

Organochlorines, 5, 165; residues, 92–93

Organophosphates (OP), 5–6, 89–92; regulation of residues, 94–95; spraying and children's exposure, 189; use in schools, 117

Oryzalin: use on school grounds, 143

Other ingredients. *See* Inert ingredients

Packaging: labeling and, 175

Paraquat, 48

Parathion, 225

Parathion methyl, 225

Pentachlorophenol, 165

Permethrin, 85

Persistence factors: pest control, 214

Persistent organic pollutants (POPs), 78, 221

Personal protective equipment, 33

Pesticide Action Network North America (PANNA), 195

Pesticide applicators: exposure, 241–42

Pesticide Data Program (PDP): U.S. Department of Agriculture (USDA), 67

Pesticide Education Center (PEL), 195

Pesticide industry: position then and now, 146–50

Pesticide manufacturers: public relations efforts, 180–81

Pesticide-related illness: complications from misdiagnosis, 41–42; data limitations, 39–40; exposure history importance, 38–39; Fresco County incident, 34–35; by industry, 29*f*; by occupation, 28*f*; by pesticide functional class, 29*f*; primary care providers and, 37–38; underreporting of, 48

Pesticides. *See also* Agricultural pesticides; Commercial pesticides; Household pesticides; Lawn and garden pesticides; Synthetic pesticides: in agriculture, 9–10, 25–63; air, water, and soil, 187–218; airborne contamination, 195–96; in the aquatic environment, 200–201; asthma and, 144–46; banned, 225–26; becoming hazardous waste, 36; blunders in the home, 168–170; the body burden and, 95, 239–240; cause, 67; caveats and uncertainty, 14; contamination of groundwater, 13; cosmetic, 236–37; daily exposures, 85, 86–87*t*; detectable residues, 67; disposal, 163–64; education, 179–180; eight fallacies about, 118–120; exposure numbers, 223; factors involved in health risks, 161–62; fate in the environment, 7; Federal regulation, 14–15; financial problems of regulation, 42–43; future prospects, 20–21; GAO study, 11–12; harmful breakdown products, 85; health effects on children, 3–4; historical patterns of use, 2–3; historical study efforts on water pollution, 202–03; homeowner awareness of pesticide risks, 163; homes, lawns, and gardens, 157–185; the human costs, 228–29; illegal application, 167–68; inadequate legal enforcement, 16–17; incidents in schools across the nation, 134–143; incomplete label data, 167–68; indoor surfaces, 160; inert ingredients, 8; information needs for registration, 16; international regulation, 220–21; international trade in, 219–233; leaching, 205; misuse of, 224; monitoring exposure, 43; myth of, 59–60; new discoveries, 4–5; pets and, 168; picture in Asia, 226–27; poisoning of children, 4; potency of, 46; product stewardship, 227; public concern, 97–98; public consciousness, 242–43; realistic considerations, 181; regulation/certification, 52–54; residues and tolerances, 17–18; resistance, 6; restricted and canceled uses, 95–97; risks to children, 162; runoff, 204; safety myths, 6; in schools, 113–155; schools and, 10–11; scope of exports, 220; small vendors, 222; soil, 208–15; State regulations, 15–16; store employees untrained regarding use, 168; three major groups, 5–6; toxicity, 7–8, 44–45; tracking water pollution, 207–08; types of produce and score, 83–84; usage, 5; using farm pesticides in the home, 161; water quality and, 12–13, 199–208; in wells, 202

Pest management: persistence factors, 214; store employees untrained regarding use, 168

Pest resistance: genetic variations and, 46–47

Pets: exposure to pesticides, 160; pesticides and, 168

Photochemical degradation, 7

Photodegradation: pesticides in soil, 212

Playground toxins, 143–44

Poisoning: of children, 4; specific cases, 229

Policy reforms, 235–36

PON-1 enzyme: pesticides and, 95–96

Posting: notification signs for pesticide applications, 177; problems with lawn postings, 178

Precautionary principle, 236

Primary care providers: pesticide issues and, 37–38

Prior Informed Consent (PIC) clause: FAO Code, 221

Spray drift, 188–89; enforcement and compliance of laws, 189; failure of laws and regulations, 199; Federal action, 190–91; incidents affecting human health, 197–98; stricter regulations, 189–190; strict liability for, 196–97

State pesticide regulations, 15–16, 35–36; schools, 126–29

Strawberries: soil fumigants and, 213–14

Sulfur compounds, 2

Surface waters: importance of, 201

Surveillance samples: food shipments, 77

Sustainable agriculture: IPM and, 238

Synthetic pesticides, 2

Telone, 195

Tenfold safety factor for children, 19

Termite control: hazards, 161

Tolerances: food safety and, 75; old, reassessment of, 76

Tomatoes: soil fumigants and, 213–14

Total Diet Study (TDS): FDA, 78

Toxicity, 7–8; the body burden, 95, 239–240; common mechanism of, 20; measurement of risk of exposure, 44–45; signal words, 45–46; tests, 82–83

Trimec, 173

Tripe-X, 165

2,4-D, 48, 164, 173

2,4,5-T, 225

U.S. Department of Agriculture (USDA): Food Safety Inspection Service (FSIS), 65–66; Pesticide Data Program (PDP), 67; reporting on organic foods, 78

U.S. Food and Drug Administration (FDA), 65–66; Total Diet Study (TDS), 78

Vapor movement, 194–95

Violative pesticide residues, 68

Washington state experience: Department of Labor and Industry's inadequate response, 44; monitoring of pesticide-related illness, 43–44; state pesticide regulations for schools, 128–29

Water quality: impact of lawn and garden pesticides, 172; pesticides and, 12–13, 199–208; prevention is key, 208; significance to, 201–02; unacceptable risks, 206–07

Wells: pesticides in, 202

Wingspread Statement, 236

Worker Protection Standard (WPS): EPA, 30–31; poor enforcement, 34; protection for children, 31–32; weakening of protection standards, 55–56

World Health Organization (WHO), 219

About the Author

MARVIN J. LEVINE is Professor of Industrial Relations (retired) at the Robert H. Smith School of Business, University of Maryland. He is the author or co-author of numerous books, including *Children for Hire: The Perils of Child Labor in the United States* (Praeger, 2003).